School Atlas
for The Commonwealth of The
BAHAMAS
for Social Studies, Tourism Education, Geography and History

T0173175

Editorial Advisor: Professor Michael Morrissey

HODDER
EDUCATION

These small maps explain the meaning of some of the lines and colours on the atlas maps.

LAND

This is how an island is shown on a map. The land is coloured green. The coastline is a blue line.

OCEANS & SEAS

There are 5 large areas of water on Earth called oceans. Smaller areas are called seas, such as the Caribbean Sea.

RIVERS & LAKES

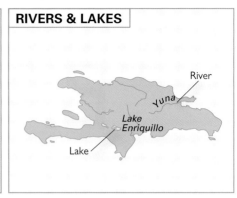

Areas of water on land are shown in a blue tint and are called lakes. Rivers are shown as blue lines.

HEIGHT OF THE LAND

The map, right, is called a topographic map. To the right below, is a diagram to show the different colours used for the height of the land. To the left below is a block diagram to give a three-dimensional impression of the map above.

Height of the land (metres)

Over 4000
2000–4000
1000–2000
400–1000
200–400
0–100
Below sea level
Sea level

COUNTRIES, CITIES & TOWNS

This is a way of showing different information about the island. It shows that the island is divided into two countries. They are separated by a country boundary. There are cities and towns on the island. The two capital cities are underlined. Large cities are shown by a red square. Other important cities are shown by a red circle.

TRANSPORT INFORMATION

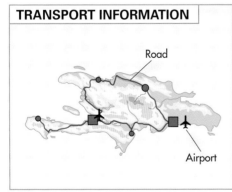

This map shows the most important roads and airports. Transport routes connect the cities and towns. Airports are shown by an aeroplane symbol.

COPYRIGHT PHILIP'S

WHERE IS THE ISLAND?

This map gives lines of latitude and longitude. These show where the island is in the world. Latitude and longitude are explained on page 5.

A COMPLETE MAP

This map is using the country colouring and shows the letter-number codes used in the index. See page 6 on how to use these codes.

Every map is designed for a specific purpose. The general maps on page 2, opposite, show for a place the height of the land and the rivers (topography). They also show the towns and cities, the roads, railways and airports. The boundaries between countries and administrative regions are shown.

Another kind of map is specially designed to show a specific topic or theme. This is a thematic map. This page has five examples of thematic maps in your atlas. Some are taken from the Caribbean Region section of the atlas, some from The Bahamas section and one from the World Data section. Read about the theme shown, look at the complete map on the correct atlas page, and then answer the questions.

MAJOR HURRICANES The theme of this map of the Caribbean is major hurricanes since 2004. Five are shown in the extract, starting in the Atlantic Ocean and moving west. The track taken by each of these hurricanes is shown with a different colour. Look at the complete map on page 13.
- Can you see some other hurricanes which began in other places?
- How many of these hurricanes passed over The Bahamas?
- Do you know names of hurricanes which are not shown on this map?

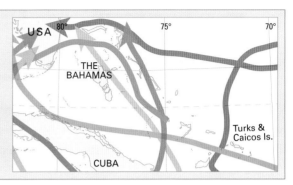

LIFE EXPECTANCY The theme of this map is how many years of life is expected for someone born in 2016. The dark green is for countries where the newly born child can expect to live for longer than average - for more than 80 years. If you live in a country shaded pale yellow, the average person will not reach 70 years of age. Look at the complete map on page 16.
- What is the life expectancy of a baby born in The Bahamas in 2016?
- Which three Caribbean countries have the lowest life expectancy?
- Why do some countries have a lower life expectancy than others?

TOURISM This extract from the map on pages 28 and 29 shows how important tourism is to The Bahamas. The map uses colours, symbols and numbers to show which islands are most important to the tourism industry. The darker pink islands have the most stopover visitors. Hotels and attractions are shown by symbols. The number beside a symbol, for example hotels, shows how many there are on each island. Look at the complete map on pages 28 and 29.
- How many islands are visited by cruise ships?

CLIMATE TYPES Each colour on this map of the world shows a type of climate. Nine climate types are shown on this map but there are many other ways to classify climates. For example, on this map yellow shows a Savanna type of climate which has high temperatures and a long dry season. Look at the complete map on page 76.
- How many regions of the world have a humid tropical climate?
- How many climate types does Australia have?
- What kind of climate does Brussels, in Europe, have?

AGRICULTURE & FISHING The map on page 30 shows where the main fishing areas are around the islands of The Bahamas – look at the darker blue colour. The main types of fish that are caught are shown in the table on the same page. Symbols show where different crops are grown. Important ferry routes are shown by the red arrows.
- How many of the fish can you recognise in the table on page 30?

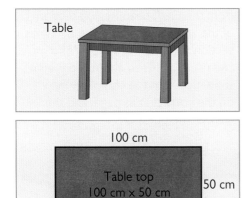

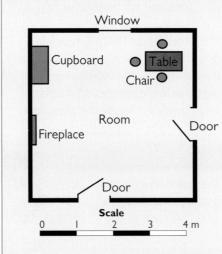

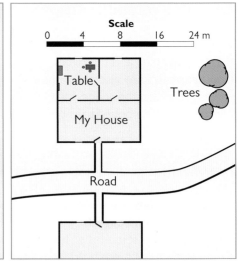

The drawing of the top of a table is looking down on it. It is 100 cm long and 50 cm wide. The drawing measures 4 x 2 cm. It is drawn to scale: 1 cm on the drawing equals 25 cm on the table.

This is a plan of a room looking down from above. 1 cm on the plan equals 1 metre in the room. The same table is shown, but now it is shown at a smaller scale.

This is an even smaller scale plan that shows the table in a room, inside a house. 1 cm on the plan equals 4 metres in the house. We can also call this a large scale map.

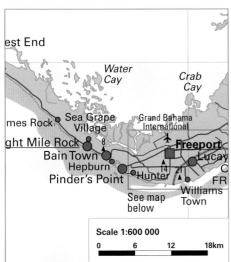

This is a map that you will find in this atlas. It is a medium-scale map and shows much more detail than the maps to its right. The map above is a larger-scale map than the maps on the right.

This is also a map from the atlas. It is at a smaller scale than the map on the left. Hence, a large-scale map shows more detail of a small area. A small-scale map shows less detail for a larger area.

This is part of the map on pages 56 and 57 in this atlas. The scale is small enough to show the whole world on two pages. This map shows the relative sizes of countries.

TYPES OF SCALE

In this atlas the scale of the map is shown in three ways.

Written Statement	**Ratio**	**Scale Bar**
This tells you how many kilometres on the earth are represented by one centimetre on the map.	This tells you that one unit on the map represents two million of the same units on the ground.	This shows you the scale as a bar. The Bahamas uses imperial as well as metric measuring systems. For the topographical maps of Bahamian islands on pages 22 to 39, miles are added to the scale bar.
1cm on the map = 20km on the ground	**Scale 1:2 000 000**	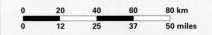

NORTH POINT

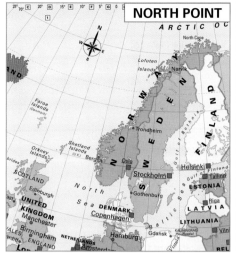

This map has a North Point showing the direction of north. It points in the same direction as the lines of longitude.

THE CARDINAL POINTS

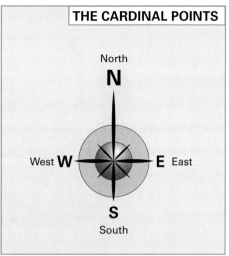

The North Point shows the four main directions: north, east, south and west. These are called the cardinal points.

THE INTERCARDINAL POINTS

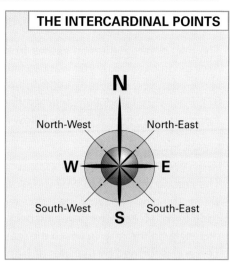

This diagram shows the intercardinal points. For example, between North and East is North-East.

LATITUDE

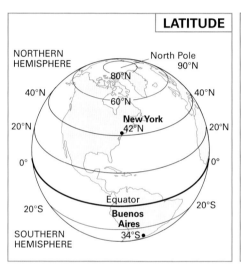

Lines of latitude cross the atlas maps from east to west. The Equator is at 0°. All other lines of latitude are either north or south of the Equator. Line 40°N is almost halfway towards the North Pole. The North Pole is at 90°N. What is the latitude of where you live?

LONGITUDE

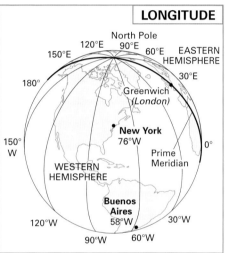

Lines of longitude run from north to south. These lines meet at the North Pole and the South Pole. Longitude 0° passes through Greenwich. This line is also called the Prime Meridian. Lines of longitude are either east of 0° or west of 0°. There are 180 degrees of longitude both east and west of 0°.

USING LATITUDE & LONGITUDE

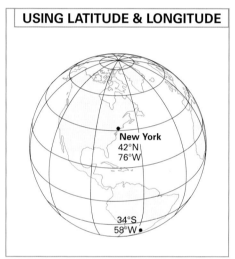

Latitude and longitude lines make a grid that can be printed on a map. You can find a place if you know its latitude and longitude. The degree of latitude is either north or south of the Equator. The longitude number is either east or west of the Greenwich Meridian.

The map on the right shows the lines of latitude which are especially important. Across the middle of the map runs the Equator which is the starting point for measuring all other lines of latitude. There are two lines called the 'Tropics'. North of the Equator, the Tropic of Cancer marks the highest latitude where the sun is overhead in the northern hemisphere. The Tropic of Capricorn marks the highest latitude where the sun is overhead in the southern hemisphere. In the North and South Polar regions, the Arctic Circle and the Antarctic Circle show the limits of the area where the Sun does not rise or set above the horizon at certain times of the year.

WHICH LINES OF LATITUDE ARE SPECIAL?

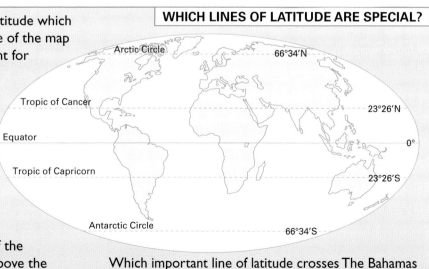

Which important line of latitude crosses The Bahamas - the Tropic of Cancer or the Tropic of Capricorn?

HOW TO FIND A PLACE

The map on the right is an extract from the map of Andros on page 36. If you want to find Andros Town in the atlas, you must look in the index (pages 79 to 81). Places are listed alphabetically. If you look up Andros Town you will find the following entry:

Andros Town **36** C2

The first number in bold type is the page number where the map appears. The letter and number code (which follow the page number) give the grid rectangle on the map in which the feature appears. The grid is formed by the lines of latitude and longitude which are shown in blue. The letter and number codes are shown in yellow boxes around the edge of the maps. Here we can see that Andros Town is on page 36 in the rectangle where column C crosses row 2.

If you need to find the latitude and longitude of a place you can work this out from a map. The latitude and longitude numbers are in black at the ends of the blue lines on the map. Latitude and longitude are measured in degrees as explained on page 5. The degree (°) is divided into 60 minutes.
   Can you work out the latitude and longitude of Andros Town?

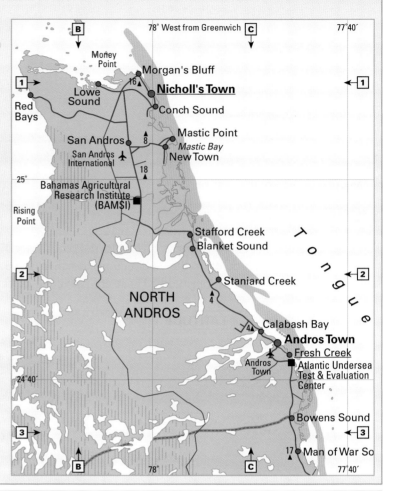

HOW TO MEASURE DISTANCE

The map on the right is a small part of the map of the Caribbean, which is on page 11 in the Caribbean Region section of the atlas.

The scale of the map extract is shown below:

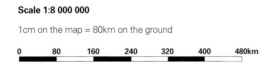

Scale 1:8 000 000

1cm on the map = 80km on the ground

| 0 | 80 | 160 | 240 | 320 | 400 | 480km |

To measure the distance from Port of Spain, Trinidad to Castries, St. Lucia you can use any of the three methods described below.

Using the written statement
Using the scale above, you can see that 1 cm on the map represents 80 km on the ground.

Measure the distance on the map between Port of Spain and Castries. You will see that it is about 4.5 cm

If 1 cm = 80 km

then 4.5 = 360 km (4.5 x 80)

Using the ratio
Using the scale above, you can see that the ratio is 1:8 000 000.

We know that the distance on the map between the cities is 4.5 cm and we know from the ratio that 1 cm on the map = 8 000 000 cm on the ground. We multiply the map distance by the ratio.

= 4.5 x 8 000 000 cm
= 36 000 000 cm
= 360 000 m
= 360 km

Using the scale bar
We know that the distance on the map between the cities is 4.5 cm.

Using the scale bar, measure 4.5 cm along this (or use a ruler as a guide) and read off the distance.

• Using these three methods, work out the distance between Port of Spain and Bridgetown, Barbados on the map above. Your teacher could tell you if your answer is correct.

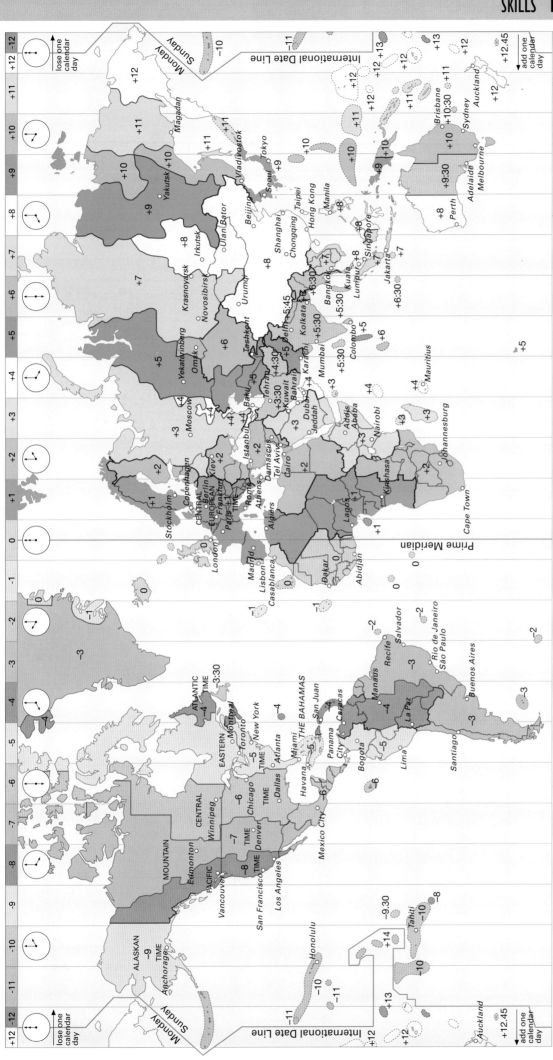

The Earth rotates on its axis. It completes one rotation in 24 hours. Because it rotates through 360°, it therefore rotates though 15° every hour. Therefore, the earth has been divided into segments of 15°, as the map shows. Each segment is a Time Zone. For each time zone, the sun is at its highest point at around mid-day.

The Prime meridian (0°) that runs from the north pole to the south pole through Greenwich in London is the standard agreed by all countries. The zone of 15° either side of the Greenwich meridian has Greenwich Mean Time or GMT. To the east, the zones are ahead of GMT (greys, greens and pale yellows on the map). To the west, the zones

are behind GMT (red and orange tints). Countries can choose which time zone to join.

The 180° line of longitude is the International Date Line. The date is different on the each side of this line.

• How many hours behind GMT is The Bahamas?

Standard Time Zones

Zones using Greenwich Mean Time
Zones behind Greenwich Mean Time
Zones ahead of Greenwich Mean Time

10 Hours slow or fast of Greenwich Mean Time
Half-hour zones
International boundaries
Time zone boundaries
International Date Line

GULF OF MEXICO

THE BAHAMAS

West Palm Beach
Fort Lauderdale
The Everglades
Miami
Freeport
Grand Bahama
Abaco
Bimini
Eleuthera Nassau
Cat I.
Great Exuma
Long I.
Acklins

Dry Tortugas (U.S.A.) Key West
Florida Keys
Florida Strait
Cay Sal Bank

CUBA
Havana Matanzas
Pinar del Río Batabanó
Santa Clara Morón
Cienfuegos
I. de la Juventud
Camagüey
Victoria de las Tunas Holguín
Guantánamo Baraco
1972 Santiago de Cuba
GUANTÁNAMO BAY (U.S.A) *Windwar*

Great Bahama Bank

Andros Island

Cayman Islands (U.K.)
George Town
Cayman Trench
Montego Bay
Savanna-la-Mar **JAMAICA**
Mandeville Kingston
Navassa I. (U.S.A)
Les Cayes

MEXICO
Progreso
Tizimín
Mérida
Cozumel *Isla Cozumel*
Yucatan Str.
C. San Antonio
C. Catoche
Cancún

Chetumal
Ambergris Cay
Belize City
Turneffe Is.
Belmopan **BELIZE**
Puerto Barrios
Gulf of Honduras
Is. de la Bahía
Roatán
La Ceiba
San Pedro Sula

HONDURAS
Central
Tegucigalpa
EL SALVADOR
San Miguel
G. de Fonseca
Chinandega
Matagalpa
NICARAGUA
León
Managua
Lake Nicaragua

Is. Santanilla (Honduras)
Pedro Cays (Jamaica)
Morant Cays (Jamaica)

Bajo Nuevo (Colombia)

C A R I B B

C. Gracias a Dios
Cayos Miskitos (Nicaragua)
Mosquito Coast
Coco (Segovia)

America

Cayos Roncador (Colombia)
I. de Providencia (Colombia)

I. de San Andrés (Colombia)
Is. del Maiz (Nicaragua)

See cross-section at bottom of page 9

Santa Marta
Barranquilla
Soledad 580
Cartagena

COSTA RICA
San José Limón
Puntarenas
G. de Nicoya
3837
Isthmus of Panama
Panama Canal
Colón
G. de los Mosquitos
Panamá
Arch. de las Perlas

Sincelejo Momp
Magangué
Montería
COLOMBIA

PACIFIC OCEAN
David
Volcan Barú 3374
Pen. de Azuero
I. de Coiba
Yaviza
Gulf of Darién
Gulf of Panama

Legend

■ Over 1,000,000 inhabitants
● Under 1,000,000 inhabitants
Panamá Capital cities underlined

COPYRIGHT PHILIP'S

—— Roads
✈ Main airports
‒‒‒ International boundaries

Swamps and marshes
Canals

Height of the land (metres)

Over 4000
2000–4000
1000–2000
400–1000
200–400
0–200
Sea level
Below sea level

Scale 1:8 000 000   1cm on the map = 80km on the ground

0   80   160   240   320   400   480km

N
W   E
S

Tropic of Cancer

A T L A N T I C

O C E A N

Mayaguana

Turks & Caicos
(U.K.)
Cockburn
Town

Inagua

assage

A T L A N T I C

O C E A N

VENEZUELA

Orinoco

Cuyuni

GUYANA

Georgetown
New Amsterdam
Bartica
Linden   Corriverton

2772 ▲ Mt.
Roraima   Kaieteur
Falls

Pakaraima Mts.

Essequibo

Corentyne

SURINAME

BRAZIL

Boa
Vista

Tacutu

Kanuku Mts.

G u i a n a

H i g h l a n d s

GUYANA
on same scale

Cap-
Haïtien
Santiago de
los Caballeros
San Francisco
de Macorís

Milwaukee
Deep
9200   *Puerto Rico*
Trench

Gonaïves
Pico Duarte
3175 ▲
DOMINICAN
REP.   La Romana

HAITI
Port-au-Prince

San
Juan

British
Virgin Is.

Anguilla
(U.K.)
St.-Martin(Fr.)

Charlotte Amalie

Barahona   **Santo**
Domingo   Ponce
PUERTO
RICO
(U.S.A.)

U.S.
Virgin Is.
St. Croix   St. Maarten
(Neth.)   St.-Barthélemy (Fr.)

ST. KITTS
& NEVIS
Saba
(Neth.)   **ANTIGUA**
& BARBUDA

H i s p a n i o l a

A n t i l l e s

Isla
Mona

Basseterre   **St. John's**

Montserrat
(U.K.)

Guadeloupe
(Fr.)
Pointe-à-Pitre

Basse-Terre

L e e w a r d   I s l a n d s

I. de Aves
(Ven.)

DOMINICA
Roseau

E A N   *S E A*

Martinique
(Fr.)
Fort-de-
France
Castries
ST. LUCIA

L e s s e r

A n t i l l e s

W i n d w a r d   I s l a n d s

Bridgetown
BARBADOS

Kingstown
ST. VINCENT
& THE
GRENADINES

St. George's   **GRENADA**

Pta. Gallinas
Oranjestad   Aruba
(Neth.)   Curaçao
(Neth.)   Bonaire (Neth.)

Pen. de la
Guajira   Pen. de
Paraguaná
Punto Fijo
Willemstad

I. Orchila
(Ven.)
Is. Los Roques
(Ven.)

I. Blanquilla
(Ven.)

Tobago

Ríohacha   Coro

I. La Tortuga
(Ven.)

I. de Margarita
Porlamar

Güiria
Port of
Spain
Arima
Trinidad

ierra Nevada de
anta Marta
Maracaibo
Valledupar   Cabimas

Puerto
Cabello   **Caracas**

San Felipe   **Maracay**

Cumaná
Carúpano   Puerto La Cruz
San
Fernando   **TRINIDAD**
& TOBAGO

Barquisimeto   **Valencia**

Barcelona

S o u t h

Maturín

Valera

Lake
Maracaibo

A m e r i c a

El Tigre

Delta of the
Orinoco

Cord. de Mérida
Barinas

V E N E Z U E L A

Ciudad
Bolívar   Ciudad Guayana

Orinoco   Boca Grande

Cúcuta   West from Greenwich   70°   San Fernando
de Apure   65°   *Orinoco*

ILLUSTRATIVE CROSS SECTION: ANDROS TO COLOMBIA

Andros
Island   Great
Bahama
Bank   Cayman Trench
(Deepest point 7680m)   Cuba   Jamaica   *Caribbean*
Sea   Colombian
Coast   2000m

Sea level
2000m
4000m
6000m
8000m

Ⓐ   Ⓑ

Cuyuni   Georgetown

GUYANA   Bartica

Linden

Pakaraima
Kaieteur
Falls   *Mts.*

2772 ▲ Mt.
Roraima 60°

COPYRIGHT PHILIP'S

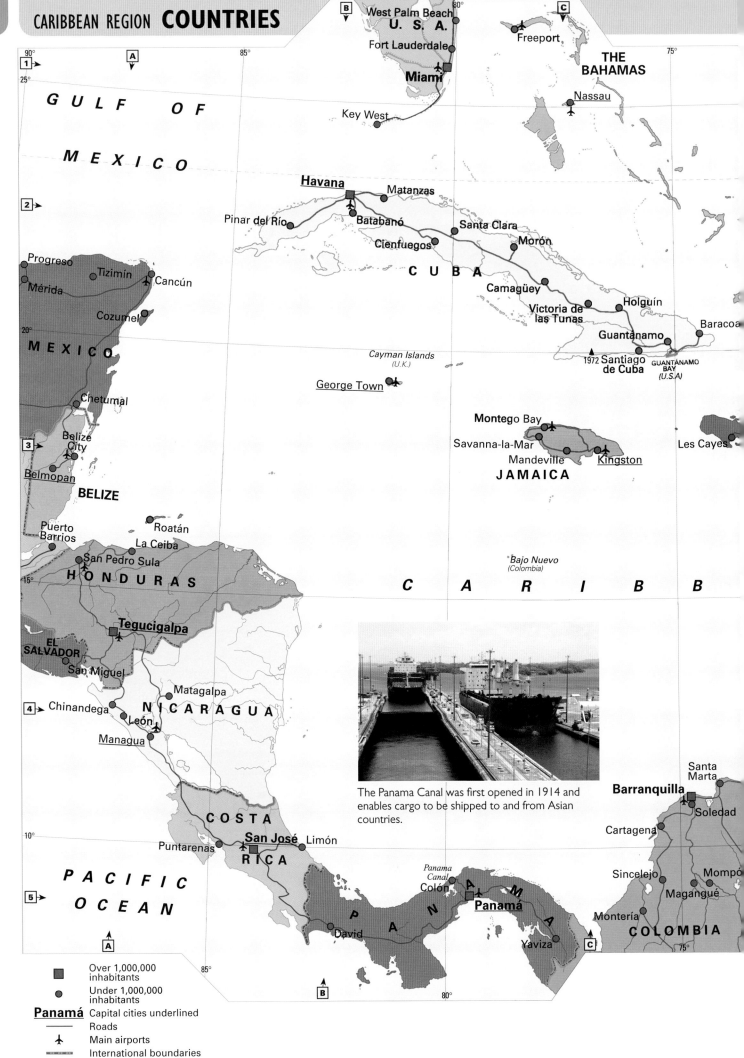

The Panama Canal was first opened in 1914 and enables cargo to be shipped to and from Asian countries.

Over 1,000,000 inhabitants
Under 1,000,000 inhabitants
Panamá Capital cities underlined
Roads
Main airports
International boundaries

COPYRIGHT PHILIP'S

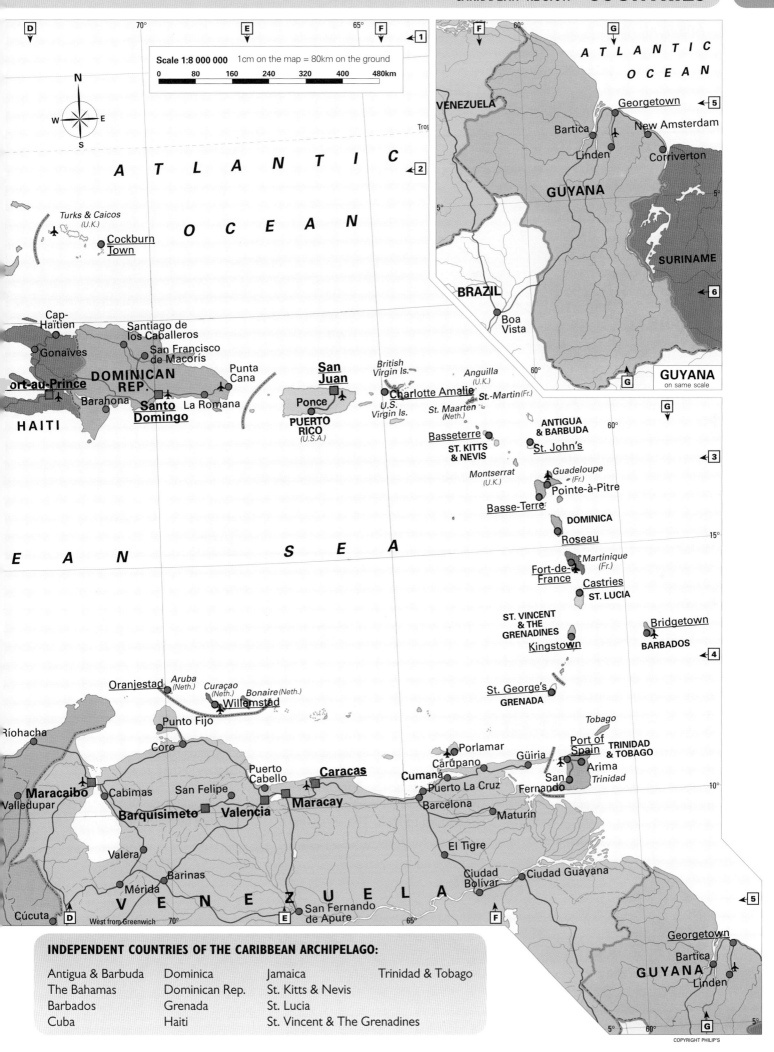

Scale 1:8 000 000   1cm on the map = 80km on the ground

0   80   160   240   320   400   480km

ATLANTIC OCEAN

Turks & Caicos (U.K.)

Cockburn Town

Cap-Haïtien

Santiago de los Caballeros

Gonaïves

San Francisco de Macorís

DOMINICAN REP.

Punta Cana

San Juan

British Virgin Is.

Anguilla (U.K.)

ort-au-Prince

Barahona

Santo Domingo

La Romana

Ponce

PUERTO RICO (U.S.A.)

Charlotte Amalie

St.-Martin (Fr.)

HAITI

U.S. Virgin Is.

St. Maarten (Neth.)

ANTIGUA & BARBUDA

Basseterre

ST. KITTS & NEVIS

St. John's

Montserrat (U.K.)

Guadeloupe (Fr.)

Pointe-à-Pitre

Basse-Terre

DOMINICA

Roseau

C A R I B B E A N   S E A

Martinique (Fr.)

Fort-de-France

Castries

ST. LUCIA

ST. VINCENT & THE GRENADINES

Bridgetown

BARBADOS

Kingstown

St. George's

GRENADA

Tobago

Aruba (Neth.)

Oranjestad

Curaçao (Neth.)

Bonaire (Neth.)

Willemstad

Port of Spain

TRINIDAD & TOBAGO

Ríohacha

Punto Fijo

Coro

Porlamar

Carúpano

Güiria

Arima

San Fernando

Trinidad

Puerto Cabello

Caracas

Cumaná

Puerto La Cruz

Maracaibo

Cabimas

San Felipe

Maracay

Barcelona

Valledupar

Barquisimeto

Valencia

Maturín

Valera

El Tigre

Barinas

V E N E Z U E L A

Ciudad Guayana

Mérida

Ciudad Bolívar

Cúcuta

West from Greenwich   70°

San Fernando de Apure

65°

Georgetown

Bartica

GUYANA

Linden

ATLANTIC OCEAN

VENEZUELA

Georgetown

Bartica

GUYANA

Linden

New Amsterdam

Corriverton

SURINAME

BRAZIL

Boa Vista

GUYANA on same scale

INDEPENDENT COUNTRIES OF THE CARIBBEAN ARCHIPELAGO:

Antigua & Barbuda	Dominica	Jamaica	Trinidad & Tobago
The Bahamas	Dominican Rep.	St. Kitts & Nevis	
Barbados	Grenada	St. Lucia	
Cuba	Haiti	St. Vincent & The Grenadines	

COPYRIGHT PHILIP'S

CARIBBEAN PLATE

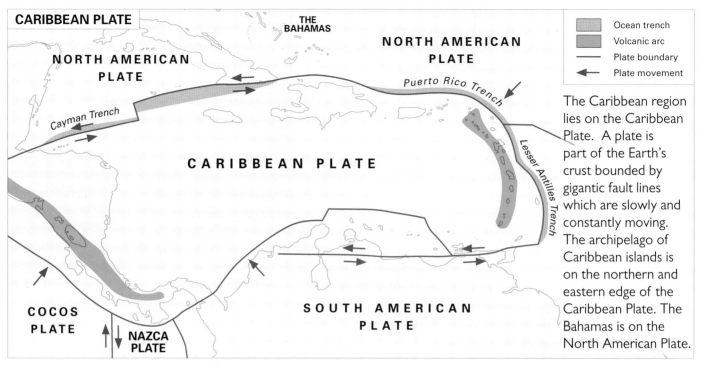

Ocean trench	
Volcanic arc	
Plate boundary	
Plate movement	

THE BAHAMAS

NORTH AMERICAN PLATE

NORTH AMERICAN PLATE

Puerto Rico Trench

Cayman Trench

CARIBBEAN PLATE

Lesser Antilles Trench

COCOS PLATE

NAZCA PLATE

SOUTH AMERICAN PLATE

The Caribbean region lies on the Caribbean Plate. A plate is part of the Earth's crust bounded by gigantic fault lines which are slowly and constantly moving. The archipelago of Caribbean islands is on the northern and eastern edge of the Caribbean Plate. The Bahamas is on the North American Plate.

EARTHQUAKES, VOLCANOES & TSUNAMIS

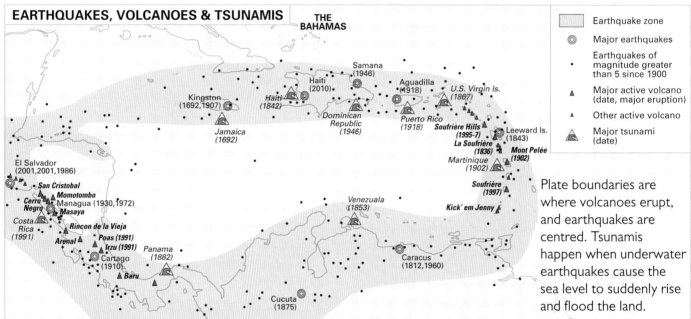

Earthquake zone	
◎	Major earthquakes
•	Earthquakes of magnitude greater than 5 since 1900
▲	Major active volcano (date, major eruption)
▴	Other active volcano
🌀	Major tsunami (date)

THE BAHAMAS

Samana (1946)
Haiti (2010)
Kingston (1692, 1907)
Haiti (1842)
Aguadilla (1918)
U.S. Virgin Is. (1867)
Dominican Republic (1946)
Puerto Rico (1918)
Soufrière Hills (1995-7)
Leeward Is. (1843)
La Soufrière (1836)
Mont Pelèe (1902)
Martinique (1902)
Jamaica (1692)
El Salvador (2001, 2001, 1986)
San Cristobal
Momotombo
Cerro Negro
Managua (1930, 1972)
Masaya
Costa Rica (1991)
Rincon de la Vieja
Arenal
Poas (1991)
Irzu (1991)
Panama (1882)
Cartago (1910)
Baru
Venezuala (1853)
Soufrière (1997)
Kick' em Jenny
Caracus (1812, 1960)
Cucuta (1875)

Plate boundaries are where volcanoes erupt, and earthquakes are centred. Tsunamis happen when underwater earthquakes cause the sea level to suddenly rise and flood the land.

PLATE TECTONICS IN THE CARIBBEAN

This cross section gives you an idea of what is happening below the surface. The line of the section runs across the Caribbean from Central America in the west to the Atlantic Ocean in the east. The North American Plate is being forced downwards under the Caribbean Plate. The Plate buckles and breaks and the rocks become hot enough to melt. The area where this happens results in the chain of volcanoes in the Leeward Islands.

- How could Bahamian islands be affected by plate movements?

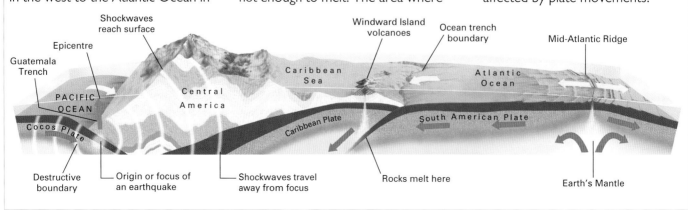

Shockwaves reach surface
Epicentre
Guatemala Trench
Windward Island volcanoes
Ocean trench boundary
Mid-Atlantic Ridge
Caribbean Sea
Atlantic Ocean
PACIFIC OCEAN
Central America
Cocos Plate
Caribbean Plate
South American Plate
Destructive boundary
Origin or focus of an earthquake
Shockwaves travel away from focus
Rocks melt here
Earth's Mantle

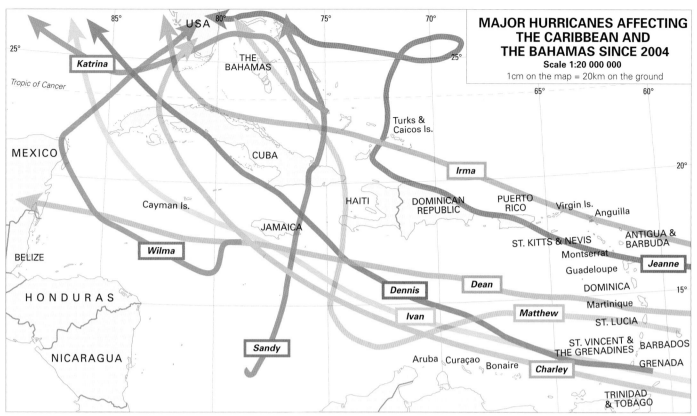

MAJOR HURRICANES AFFECTING THE CARIBBEAN AND THE BAHAMAS SINCE 2004
Scale 1:20 000 000
1cm on the map = 20km on the ground

Name	Places badly affected	Period	Maximum strength	Total deaths
Charley	Barbados, Cayman Is., Jamaica, Cuba	9–14 Aug. 2004	4	10
Ivan	Tobago, Grenada, Jamaica, Cayman Is., Cuba	20–24 Sept. 2004	5	92
Jeanne	Guadeloupe, Puerto Rico, Dominican Republic, Haiti, Bahamas	13–28 Sept. 2004	3	3000+
Dennis	Cuba	4–18 Jul. 2005	4	88
Katrina	Bahamas	28–29 Aug. 2005	5	1,836
Wilma	Haiti, Jamaica	19 Oct. 2005	5	87
Dean	Lesser Antilles, Jamaica	18–21 Aug. 2007	5	45
Sandy	Jamaica, Cuba, Bahamas	22–29 Oct. 2012	3	286
Matthew	Haiti, Cuba, Dominican Republic, Bahamas	30 Sept.–7 Oct. 2016	5	603
Irma	Barbuda, St. Martin, Anguilla, Virgin Is., Turks & Caicos,	30 Aug.–16 Sept. 2017	5	102

SAFFIR-SIMPSON HURRICANE SCALE

Category 1 – Weak hurricane
Winds 119–153 km/hour. Some damage and power cuts.

Category 2 – Moderate hurricane
Winds 154–177 km/hour. Extensive damage and power cuts. Many trees uprooted or snapped.

Category 3 – Strong hurricane
Winds 178–208 km/hour. Well-built homes suffer major damage. Trees uprooted. Flooding inland. Total power loss.

Category 4 – Very strong hurricane
Winds 209–251 km/hour. Catastrophic damage. Severe damage to well-built homes, trees blown over.

Category 5 – Devastating hurricane
Winds over 252 km/hour. Many buildings destroyed, major roads cut off. Damage by storm surge.

CROSS-SECTION THROUGH A HURRICANE

Total width 200–800 km

Cirrus cloud

Dense cloud

Thunderstorms

Height 12 km

Cooled air spirals outwards and descends

Vortex of hurricane

Warm, moist air spirals towards and around the eye of the hurricane, rising and cooling rapidly

Gusty winds

Violent winds (250 km/h)

Calm eye

Westerly path of system

Energy from warm sea (over 27°C)

Hispaniola

South America

Hurricane Irma (above), with winds of 295 km/hour, was the most powerful in over ten years when it made landfall on Barbuda in September 2017. It caused catastrophic damage in St. Barthélemy, St. Martin, Anguilla and the Virgin Islands.

ERA OF EUROPEAN EXPLORATION

The Ciboneys, Arawaks and Caribs (Tainos and Kalinagos) settled the Caribbean long before the arrival of Europeans. It is estimated that at the time of Columbus' voyages there were 200,000 Amerindian peoples in the Caribbean region. Many were killed by the explorers and many more by diseases which the Europeans introduced.

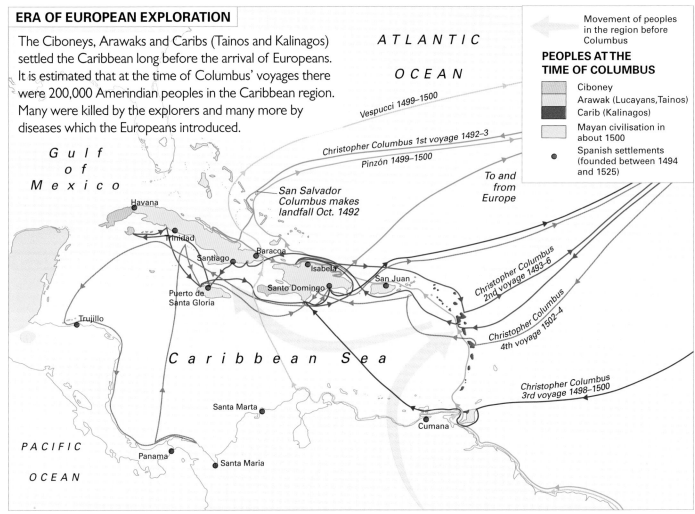

Movement of peoples in the region before Columbus

PEOPLES AT THE TIME OF COLUMBUS

Ciboney
Arawak (Lucayans, Tainos)
Carib (Kalinagos)
Mayan civilisation in about 1500
● Spanish settlements (founded between 1494 and 1525)

AGE OF SPANISH CONQUEST OF AMERICAN CIVILISATIONS

The Spaniards were mainly interested in the gold and silver of South America. They also colonised islands of the Greater Antilles. Other European countries began to challenge the Spaniards in the Caribbean and to occupy islands in the Lesser Antilles. They cultivated tobacco, indigo, sugar and other crops for export.

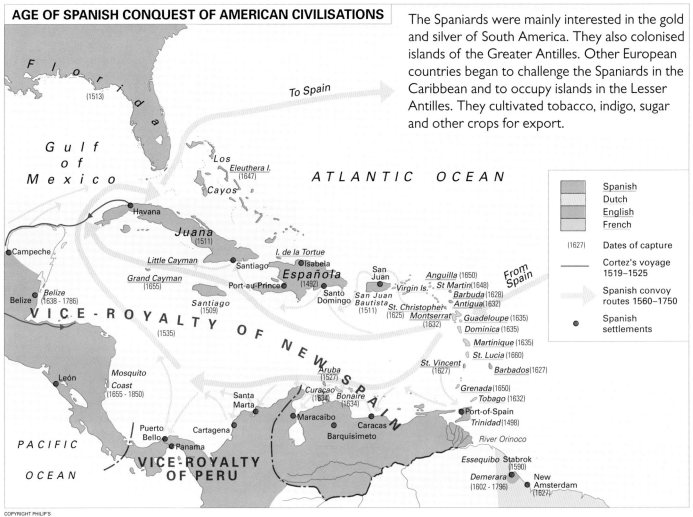

Spanish
Dutch
English
French

(1627) Dates of capture
— Cortez's voyage 1519–1525
⇒ Spanish convoy routes 1560–1750
● Spanish settlements

COLONIES

	Spanish
	Dutch
	British
	French
	U.S.

(1638) Dates of acquisition

⟷ Slave routes and indentured labour routes

✦ Rebellions

● Major towns and naval bases

——— Today's international boundaries

EUROPEAN SLAVE TRADE & AFRICAN REBELLION

European demand for sugar grew. Over four million Africans were enslaved and shipped to the West Indies. In all, over ten million Africans were taken across the Atlantic in slave ships.

• What work were enslaved Africans forced to do in Bahamian islands?

• When was slavery prohibited in your country?

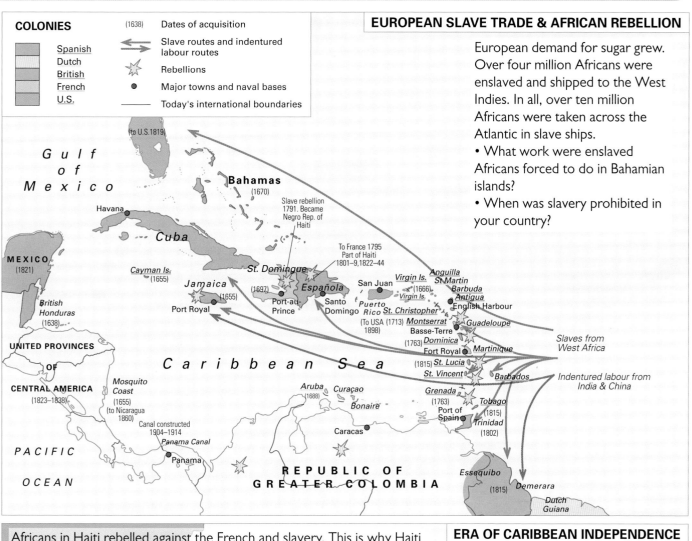

ERA OF CARIBBEAN INDEPENDENCE

Africans in Haiti rebelled against the French and slavery. This is why Haiti was the first to be independent. There was much emigration to Europe and America after 1950. When did The Bahamas gain its independence? Which territory southeast of The Bahamas is not yet independent?

1821	Date of independence
✦	U.S. intervention in Caribbean territories
←	Emigration from the West Indies
———	Today's international boundaries

ISLANDS NOT INDEPENDENT

	Dutch
	British
	French
	U.S.

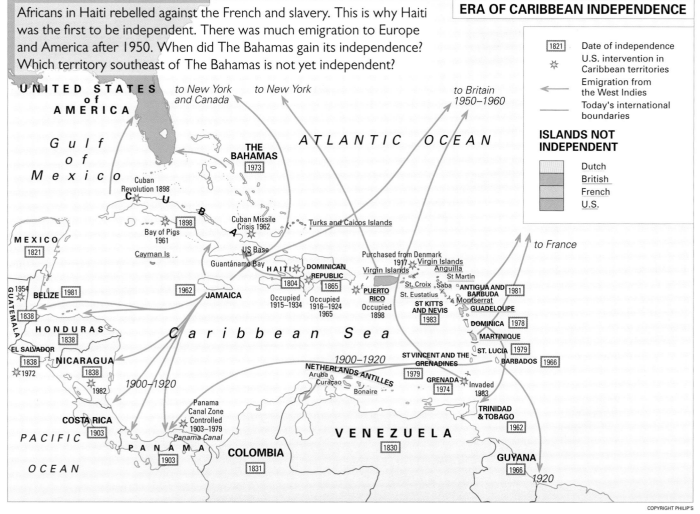

COPYRIGHT PHILIP'S

DENSITY of POPULATION

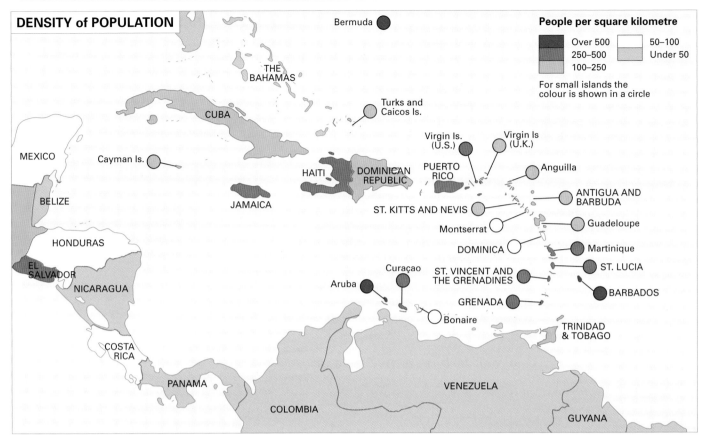

People per square kilometre

Over 500	50–100
250–500	Under 50
100–250	

For small islands the colour is shown in a circle

Bermuda

THE BAHAMAS

CUBA

Turks and Caicos Is.

MEXICO

Cayman Is.

HAITI

DOMINICAN REPUBLIC

PUERTO RICO

Virgin Is. (U.S.)

Virgin Is (U.K.)

Anguilla

BELIZE

JAMAICA

ST. KITTS AND NEVIS

ANTIGUA AND BARBUDA

HONDURAS

Montserrat

Guadeloupe

DOMINICA

Martinique

EL SALVADOR

NICARAGUA

Curaçao

ST. VINCENT AND THE GRENADINES

ST. LUCIA

Aruba

Bonaire

GRENADA

BARBADOS

COSTA RICA

TRINIDAD & TOBAGO

PANAMA

VENEZUELA

COLOMBIA

GUYANA

The Caribbean region has both areas of high and low population density. The Bahamas is a country with a small population. Nassau on New Providence (left) has a high population density with many buildings and with people living close together. The Exuma islands (right) are typical of areas of low population density.

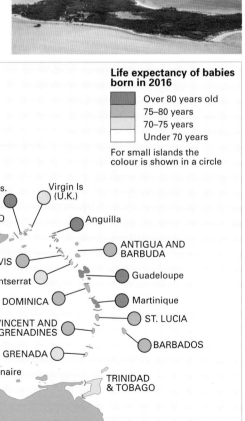

LIFE EXPECTANCY

Life expectancy of babies born in 2016

	Over 80 years old
	75–80 years
	70–75 years
	Under 70 years

For small islands the colour is shown in a circle

Bermuda

THE BAHAMAS

CUBA

Turks and Caicos Is.

MEXICO

Cayman Is.

HAITI

DOMINICAN REPUBLIC

PUERTO RICO

Virgin Is. (U.S.)

Virgin Is (U.K.)

Anguilla

BELIZE

JAMAICA

ST. KITTS AND NEVIS

ANTIGUA AND BARBUDA

HONDURAS

Montserrat

Guadeloupe

DOMINICA

Martinique

EL SALVADOR

NICARAGUA

Curaçao

ST. VINCENT AND THE GRENADINES

ST. LUCIA

Aruba

Bonaire

GRENADA

BARBADOS

COSTA RICA

TRINIDAD & TOBAGO

PANAMA

VENEZUELA

COLOMBIA

GUYANA

MIGRATION

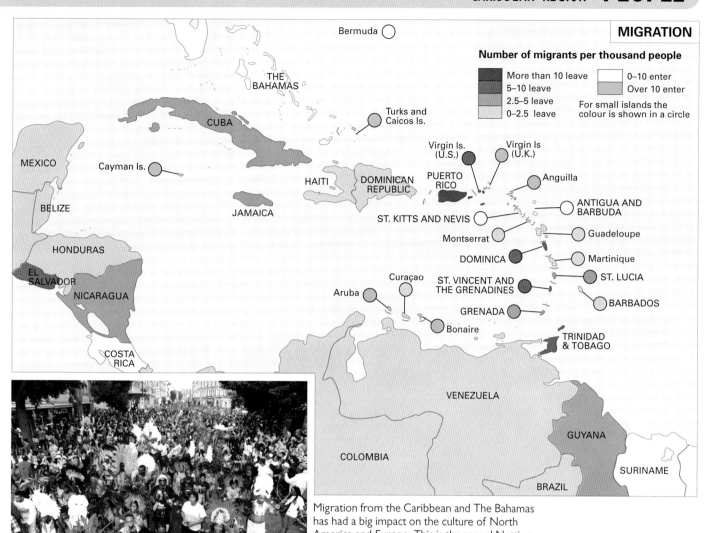

Bermuda

THE BAHAMAS

CUBA

MEXICO

Cayman Is.

BELIZE

JAMAICA

HAITI

DOMINICAN REPUBLIC

PUERTO RICO

HONDURAS

EL SALVADOR

NICARAGUA

COSTA RICA

Turks and Caicos Is.

Virgin Is. (U.S.)

Virgin Is (U.K.)

Anguilla

ST. KITTS AND NEVIS

ANTIGUA AND BARBUDA

Montserrat

Guadeloupe

DOMINICA

Martinique

Curaçao

ST. VINCENT AND THE GRENADINES

ST. LUCIA

Aruba

Bonaire

GRENADA

BARBADOS

TRINIDAD & TOBAGO

VENEZUELA

GUYANA

COLOMBIA

SURINAME

BRAZIL

Number of migrants per thousand people

- More than 10 leave
- 5–10 leave
- 2.5–5 leave
- 0–2.5 leave
- 0–10 enter
- Over 10 enter

For small islands the colour is shown in a circle

Migration from the Caribbean and The Bahamas has had a big impact on the culture of North America and Europe. This is the annual Notting Hill carnival in London.

LANGUAGES

Bermuda

THE BAHAMAS

CUBA

MEXICO

Cayman Is.

BELIZE

HAITI

DOMINICAN REPUBLIC

PUERTO RICO

JAMAICA

HONDURAS

EL SALVADOR

NICARAGUA

COSTA RICA

PANAMA

Turks and Caicos Is.

Virgin Is. (U.S.)

Virgin Is (U.K.)

Anguilla

ST. KITTS AND NEVIS

ANTIGUA AND BARBUDA

Montserrat

Guadeloupe

DOMINICA

Martinique

Curaçao

ST. VINCENT AND THE GRENADINES

ST. LUCIA

Aruba

Bonaire

GRENADA

BARBADOS

TRINIDAD & TOBAGO

VENEZUELA

GUYANA

COLOMBIA

SURINAME

BRAZIL

Official Languages

- English
- Spanish
- French
- Dutch
- Portuguese

For small islands the colour is shown in a circle

In many countries the people use creole languages alongside the official national language, such as Papiamento in Curaçao, French Creole in Dominica and Patois in Jamaica.

COPYRIGHT PHILIP'S

Citrus farms in Belize produce oranges on a large scale for processing to juice.

Jamaica's Blue Mountain coffee beans are handpicked to preserve quality.

Jamaican bananas for the export market have to be carefully prepared before shipping.

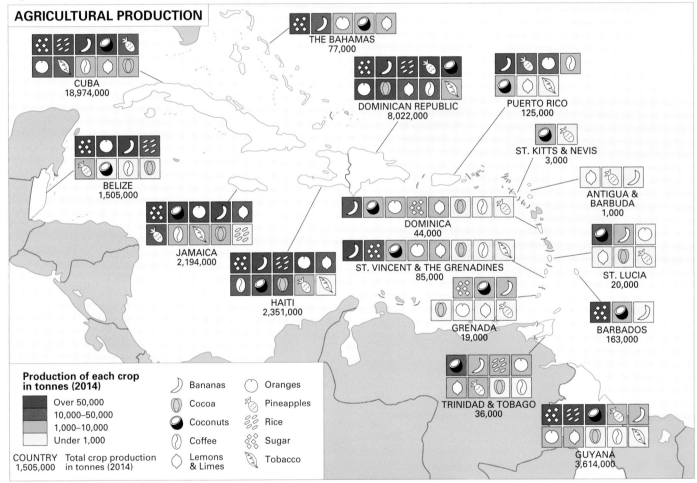

AGRICULTURAL PRODUCTION

THE BAHAMAS
77,000

CUBA
18,974,000

DOMINICAN REPUBLIC
8,022,000

PUERTO RICO
125,000

ST. KITTS & NEVIS
3,000

BELIZE
1,505,000

ANTIGUA &
BARBUDA
1,000

JAMAICA
2,194,000

DOMINICA
44,000

ST. LUCIA
20,000

HAITI
2,351,000

ST. VINCENT & THE GRENADINES
85,000

GRENADA
19,000

BARBADOS
163,000

TRINIDAD & TOBAGO
36,000

GUYANA
3,614,000

Production of each crop in tonnes (2014)

- Over 50,000
- 10,000–50,000
- 1,000–10,000
- Under 1,000

COUNTRY   Total crop production
1,505,000   in tonnes (2014)

- Bananas
- Cocoa
- Coconuts
- Coffee
- Lemons & Limes
- Oranges
- Pineapples
- Rice
- Sugar
- Tobacco

Sugar cane is increasingly harvested by machine, as it is in St. Kitts.

Grenada is the world's second largest producer of nutmeg. Nutmeg products are exported.

Machines are now used to harvest the rice crop along Guyana's flat coastal lands.

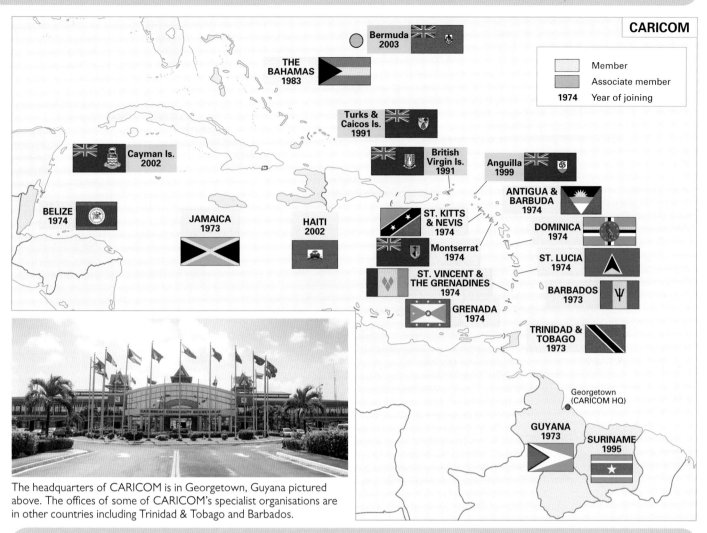

CARICOM

	Member
	Associate member
1974	Year of joining

Bermuda 2003

THE BAHAMAS 1983

Turks & Caicos Is. 1991

Cayman Is. 2002

British Virgin Is. 1991

Anguilla 1999

ANTIGUA & BARBUDA 1974

BELIZE 1974

JAMAICA 1973

HAITI 2002

ST. KITTS & NEVIS 1974

DOMINICA 1974

Montserrat 1974

ST. LUCIA 1974

ST. VINCENT & THE GRENADINES 1974

BARBADOS 1973

GRENADA 1974

TRINIDAD & TOBAGO 1973

Georgetown (CARICOM HQ)

GUYANA 1973

SURINAME 1995

The headquarters of CARICOM is in Georgetown, Guyana pictured above. The offices of some of CARICOM's specialist organisations are in other countries including Trinidad & Tobago and Barbados.

ORGANISATIONS OF THE CARIBBEAN COMMUNITY

Caribbean Agricultural Development Institute – CARDI
Caribbean Agricultural Health and Safety Agency – CAHFSA
Caribbean Aviation Safety and Security Oversight System – CASSOS
Caribbean Centre for Renewable Energy and Energy Efficiency – CCREEE
Caribbean Centre for Development Administration – CARICAD
Caribbean Community Climate Change Centre – CCCCC
Caribbean Court of Justice – CJC
Caribbean Examinations Council – CXC
Caribbean Institute for Meteorology and Hydrology – CIMHH
Caribbean Meteorological Organisation – CMO

Caribbean Regional Fisheries Mechanism – CRFM
Caribbean Telecommunications Union – CTU
Caribbean Competition Commission – CCC
Caribbean Development Fund – CDF
Caribbean Implementing Agency for Crime and Security – IMPACS

Examples of associate institutions

Caribbean Development Bank – CDB
Caribbean Disaster Emergency Management Agency – CDEMA
Caribbean Export Development Agency – Carib-Export
Caribbean Law Institute – CLI
Caribbean Tourism Organisation – CTO

The Caribbean Community has established many organisations to improve the region's economy, safety and justice. These are listed above. Other regional organisations are associated with CARICOM. Some examples of these are listed.

- Using the name for each organisation, can you work out how it will help the people of the Caribbean?
- Most organisations have a logo to give it identity. Can you match the logos below with one of the organisations listed?

CARICOM

The Caribbean Community (CARICOM) began in 1973 when Barbados, Guyana, Jamaica and Trinidad and Tobago signed the Treaty of Chaguaramas (in Trinidad). The other members joined in 1974 and at later dates with Haiti the last to join in 2002. The aims are: economic cooperation within the Caribbean Common Market, coordination of foreign policy and cooperation in the areas of health, education, technology, transport, culture and sport. The headquarters is in Guyana.

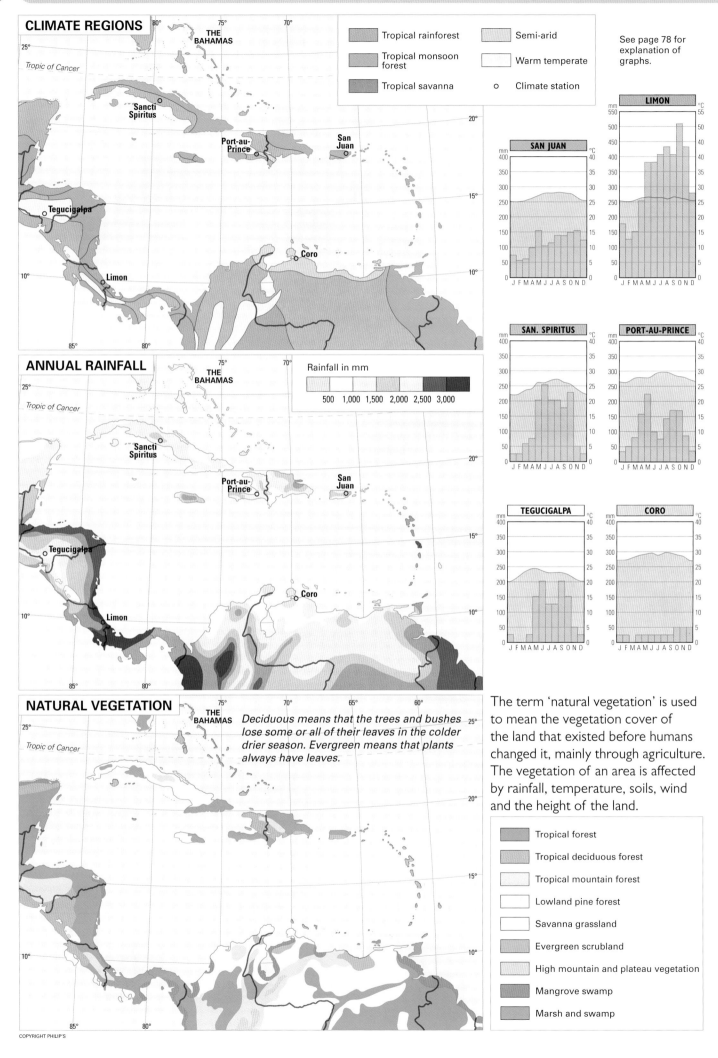

CLIMATE REGIONS

- Tropical rainforest
- Tropical monsoon forest
- Tropical savanna
- Semi-arid
- Warm temperate
- o Climate station

See page 78 for explanation of graphs.

THE BAHAMAS

Tropic of Cancer

Sancti Spiritus

Port-au-Prince

San Juan

Tegucigalpa

Coro

Limon

ANNUAL RAINFALL

Rainfall in mm

500 1,000 1,500 2,000 2,500 3,000

THE BAHAMAS

Tropic of Cancer

Sancti Spiritus

Port-au-Prince

San Juan

Tegucigalpa

Coro

Limon

NATURAL VEGETATION

THE BAHAMAS

Deciduous means that the trees and bushes lose some or all of their leaves in the colder drier season. Evergreen means that plants always have leaves.

Tropic of Cancer

The term 'natural vegetation' is used to mean the vegetation cover of the land that existed before humans changed it, mainly through agriculture. The vegetation of an area is affected by rainfall, temperature, soils, wind and the height of the land.

- Tropical forest
- Tropical deciduous forest
- Tropical mountain forest
- Lowland pine forest
- Savanna grassland
- Evergreen scrubland
- High mountain and plateau vegetation
- Mangrove swamp
- Marsh and swamp

SAN JUAN

LIMON

SAN. SPIRITUS

PORT-AU-PRINCE

TEGUCIGALPA

CORO

COPYRIGHT PHILIP'S

TOURIST ARRIVALS

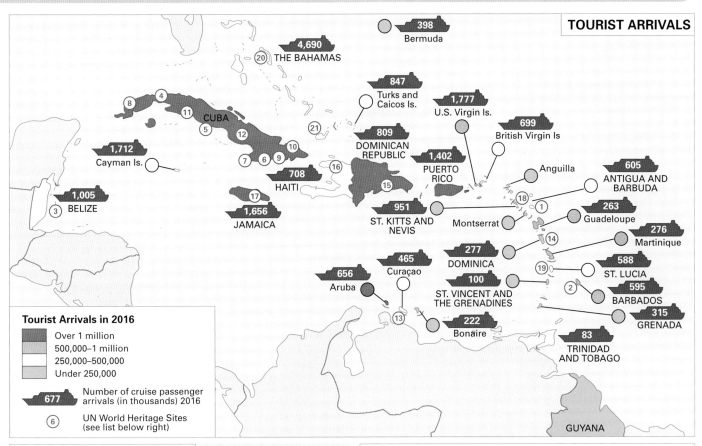

Bermuda 398

THE BAHAMAS 4,690
20

Turks and Caicos Is. 847

U.S. Virgin Is. 1,777

British Virgin Is 699

Cayman Is. 1,712

DOMINICAN REPUBLIC 809

PUERTO RICO 1,402

Anguilla

ANTIGUA AND BARBUDA 605

BELIZE 1,005
3

HAITI 708

JAMAICA 1,656

ST. KITTS AND NEVIS 951

Montserrat

Guadeloupe 263

Martinique 276

DOMINICA 277

ST. LUCIA 588

Curaçao 465

ST. VINCENT AND THE GRENADINES 100

BARBADOS 595

GRENADA 315

Aruba 656

Bonaire 222

TRINIDAD AND TOBAGO 83

GUYANA

Tourist Arrivals in 2016

- Over 1 million
- 500,000–1 million
- 250,000–500,000
- Under 250,000

677 Number of cruise passenger arrivals (in thousands) 2016

6 UN World Heritage Sites (see list below right)

EMPLOYMENT IN TOURISM

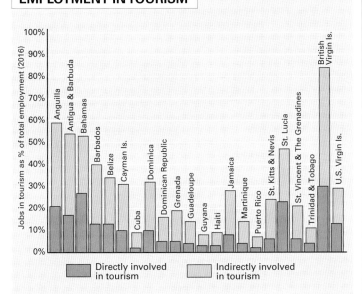

Jobs in tourism as % of total employment (2016)

■ Directly involved in tourism
□ Indirectly involved in tourism

UN WORLD HERITAGE SITES IN THE CARIBBEAN & PROPOSED FOR THE BAHAMAS

1. Antigua Naval Dockyard, Antigua & Barbuda
2. Historic Bridgetown and its Garrison, Barbados
3. Belize Barrier Reef Reserve System, Belize
4. Old Havana and its Fortifications, Cuba
5. Trinidad and the Valley de los Ingenios, Cuba
6. San Pedro de la Roca Castle, Santiago de Cuba, Cuba
7. Desembarco del Granma National Park, Cuba
8. Viñales Valley, Cuba
9. Landscapes of the First Coffee Plantations, Cuba
10. Alejandro de Humboldt National Park, Cuba
11. Historic Centre of Cienfuegos, Cuba
12. Historic Centre of Camagüey, Cuba
13. Historic Area of Willemstad, Inner City and Harbour, Curaçao
14. Morne Trois Pitons National Park, Dominica
15. Colonial City of Santo Domingo, Dominican Republic
16. National History Park, Sans Souci Citadel, Haiti
17. Blue and John Crow Mountains, Jamaica
18. Brimstone Hill Fortress National Park, St. Kitts & Nevis
19. Pitons Management Area, St. Lucia
    Nominated:
20. Historic Lighthouses of The Bahamas (proposed)
21. The Inagua National Park, The Bahamas (proposed)

IMPORTANCE OF TOURISM

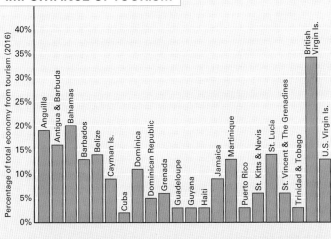

Percentage of total economy from tourism (2016)

Selected by UNESCO, World Heritage sites are places of great international importance. The Pitons in St. Lucia (two volcanic mountains) are a good example of such a site.

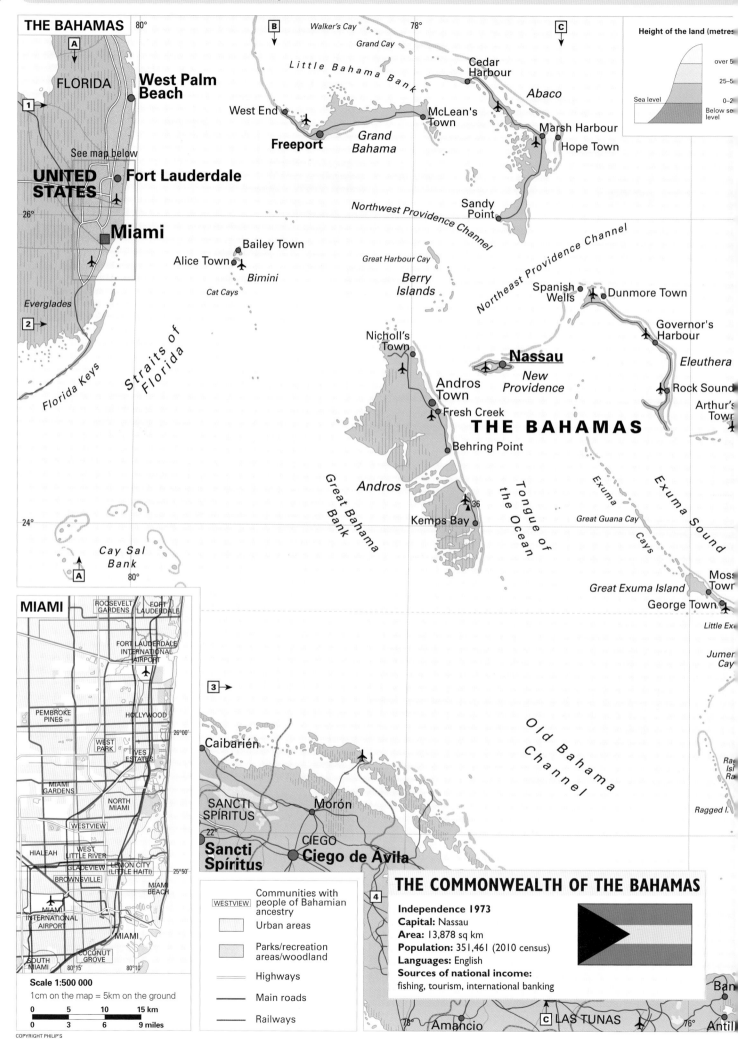

THE BAHAMAS   80°

A

FLORIDA

West Palm Beach

West End   McLean's Town

Freeport   *Grand Bahama*

Walker's Cay   Grand Cay

Little Bahama Bank

Cedar Harbour

Abaco

Marsh Harbour

Hope Town

Sandy Point

See map below

UNITED STATES   **Fort Lauderdale**

Miami

Bailey Town

Alice Town

Bimini

Cat Cays

Everglades

26°

Great Harbour Cay

Berry Islands

Spanish Wells

Dunmore Town

Governor's Harbour

Eleuthera

Rock Sound

Arthur's Town

Nicholl's Town

Nassau

New Providence

Andros Town

Fresh Creek

THE BAHAMAS

Behring Point

Florida Keys

Straits of Florida

Great Bahama Bank

Andros

Kemps Bay

36

Tongue of the Ocean

Exuma Cays

Exuma Sound

Great Guana Cay

24°

Cay Sal Bank
80°

A

Moss Town

Great Exuma Island

George Town

Little Ex

Jumer Cay

Northwest Providence Channel

Northeast Providence Channel

Height of the land (metres

over 5

25–5

0–2

Sea level

Below se level

MIAMI

ROOSEVELT GARDENS   FORT LAUDERDALE

FORT LAUDERDALE INTERNATIONAL AIRPORT

PEMBROKE PINES   HOLLYWOOD

26°00'

WEST PARK   IVES ESTATES

MIAMI GARDENS

NORTH MIAMI

WESTVIEW

HIALEAH   WEST LITTLE RIVER

LEMON CITY (LITTLE HAITI)

GLADEVIEW

BROWNSVILLE

MIAMI BEACH

25°50'

MIAMI INTERNATIONAL AIRPORT

MIAMI

COCONUT GROVE

SOUTH MIAMI
80°15'   80°10'

Scale 1:500 000

1cm on the map = 5km on the ground

| 0 | 5 | 10 | 15 km |

| 0 | 3 | 6 | 9 miles |

COPYRIGHT PHILIP'S

3

Caibarién

SANCTI SPÍRITUS

Morón

Sancti Spíritus

CIEGO

Ciego de Ávila

22°

Old Bahama Channel

Ra Isl Ra

Ragged I.

4

Amancio   C LAS TUNAS   Antil

Ban

78°   76°

WESTVIEW	Communities with people of Bahamian ancestry
	Urban areas
	Parks/recreation areas/woodland
	Highways
	Main roads
	Railways

THE COMMONWEALTH OF THE BAHAMAS

Independence 1973
Capital: Nassau
Area: 13,878 sq km
Population: 351,461 (2010 census)
Languages: English
Sources of national income:
fishing, tourism, international banking

Urban areas
Towns and villages
ssau Capital city underlined
Highways
Roads
Airports
International boundaries
Mangroves   Reefs

TURKS & CAICOS ISLANDS

G **H**

22°

Caicos Passage
North Caicos
Bottle Creek

Providenciales I.

Blue Hills
Conch Bar
Bambarra

Middle Caicos

West Caicos I.

C a i c o s   I s l a n d s

East Caicos

35 ▲

South Caicos

26°

Caicos Bank

Cockburn Harbour

Turks Island Passage

A T L A N T I C   O C E A N

Cockburn Town

Grand Turk I.

1

Salt Cay

Turks Islands

1

Ambergris Cays

Seal Cays

West 72° from Greenwich

G **H**

TURKS & CAICOS ISLANDS

British Overseas Territory
Capital: Cockburn Town
Area: 430 sq km
Population: 40,000 (est. 2018)
Languages: English (official)
Sources of national income: tourism, financial services, fishing

Scale 1:1 240 000
1cm on the map = 12.4km on the ground

| 0 | 12.4 | 24.8 | 37.2 | 49.6 km |
| 0 | 7.7 | 15.4 | 23.1 | 30.8 miles |

74° **E** 72° **F**

N
W — E
S

Island

New Bight

San Salvador I.
Cockburn Town

Rum Cay

Long Island   Tropic of Cancer

24°

Deadman's Cay

Samana Cay

Clarence Town

Crooked Island Passage

Crooked Island

Colonel Hill

Plana Cays

Albert Town

Mayaguana Passage

Mayaguana
Abraham's Bay

Acklins

Caicos Passage

A T L A N T I C
O C E A N

See map above

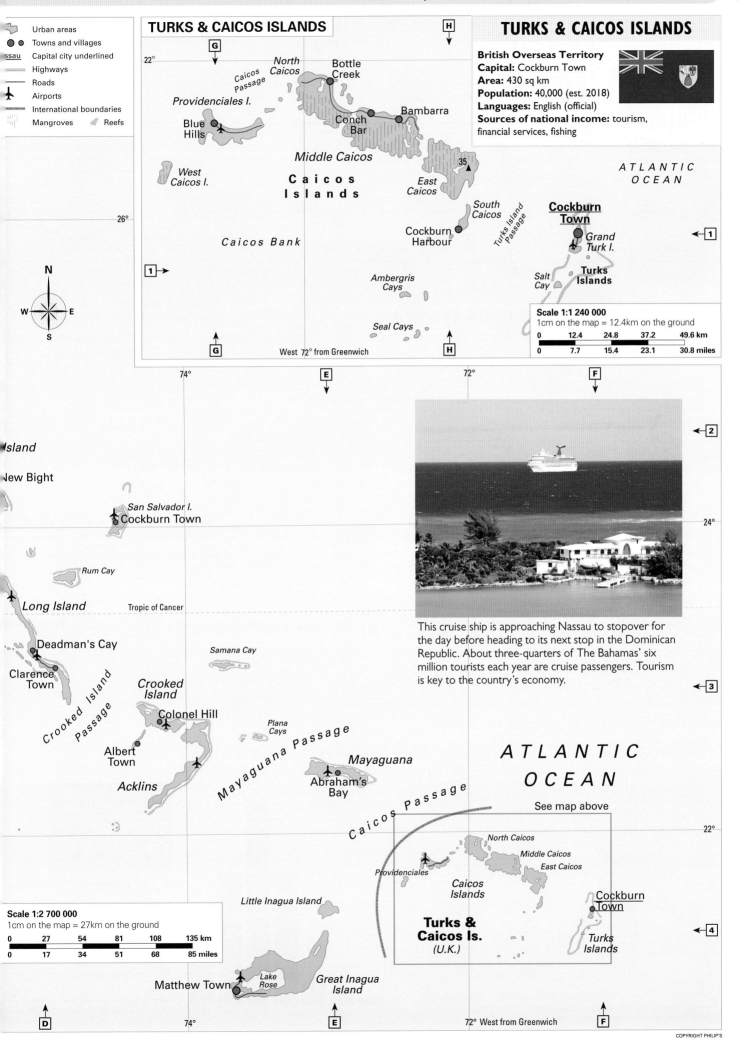

This cruise ship is approaching Nassau to stopover for the day before heading to its next stop in the Dominican Republic. About three-quarters of The Bahamas' six million tourists each year are cruise passengers. Tourism is key to the country's economy.

22°

North Caicos
Middle Caicos
East Caicos

Providenciales

Caicos Islands

Turks & Caicos Is.
(U.K.)

Cockburn Town

Turks Islands

Scale 1:2 700 000
1cm on the map = 27km on the ground

| 0 | 27 | 54 | 81 | 108 | 135 km |
| 0 | 17 | 34 | 51 | 68 | 85 miles |

Little Inagua Island

Matthew Town

Lake Rose

Great Inagua Island

D 74° **E** 72° West from Greenwich **F**

2

3

4

RAINFALL & TEMPERATURE

Rainfall in mm	
	Over 1250
	1000–1250
	750–1000
	Under 750

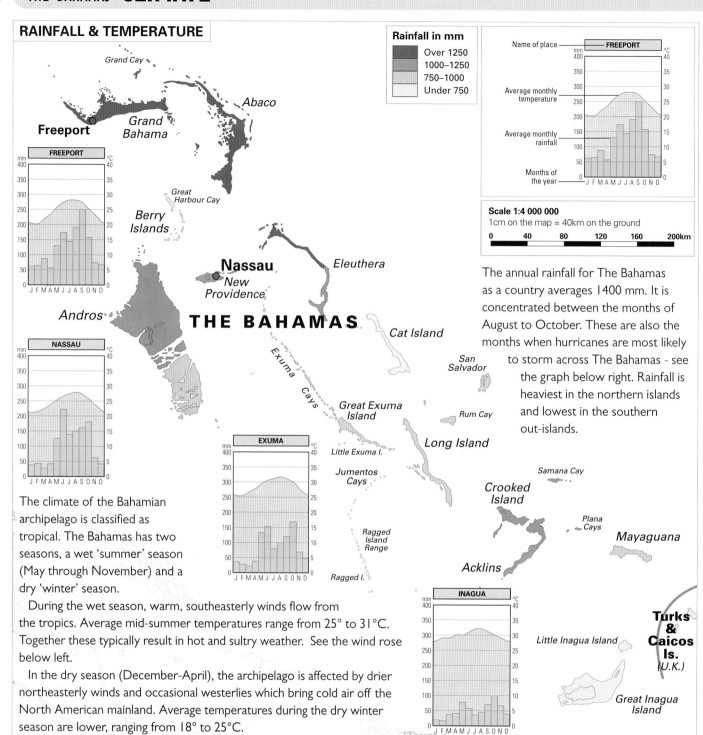

Scale 1:4 000 000
1cm on the map = 40km on the ground

0 40 80 120 160 200km

The annual rainfall for The Bahamas as a country averages 1400 mm. It is concentrated between the months of August to October. These are also the months when hurricanes are most likely to storm across The Bahamas - see the graph below right. Rainfall is heaviest in the northern islands and lowest in the southern out-islands.

The climate of the Bahamian archipelago is classified as tropical. The Bahamas has two seasons, a wet 'summer' season (May through November) and a dry 'winter' season.

During the wet season, warm, southeasterly winds flow from the tropics. Average mid-summer temperatures range from 25° to 31°C. Together these typically result in hot and sultry weather. See the wind rose below left.

In the dry season (December-April), the archipelago is affected by drier northeasterly winds and occasional westerlies which bring cold air off the North American mainland. Average temperatures during the dry winter season are lower, ranging from 18° to 25°C.

WIND DIRECTION

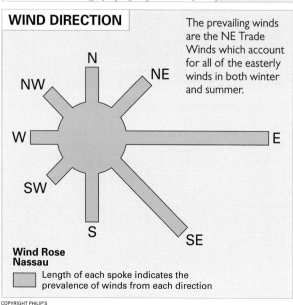

The prevailing winds are the NE Trade Winds which account for all of the easterly winds in both winter and summer.

Wind Rose Nassau
Length of each spoke indicates the prevalence of winds from each direction

HURRICANE FREQUENCY & NAMES

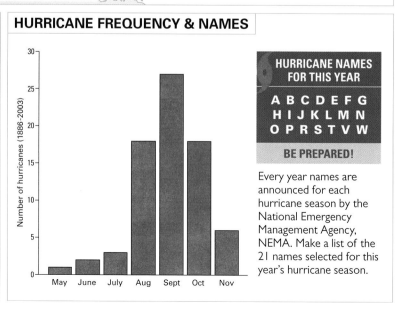

HURRICANE NAMES FOR THIS YEAR

A B C D E F G
H I J K L M N
O P R S T V W

BE PREPARED!

Every year names are announced for each hurricane season by the National Emergency Management Agency, NEMA. Make a list of the 21 names selected for this year's hurricane season.

ENVIRONMENT

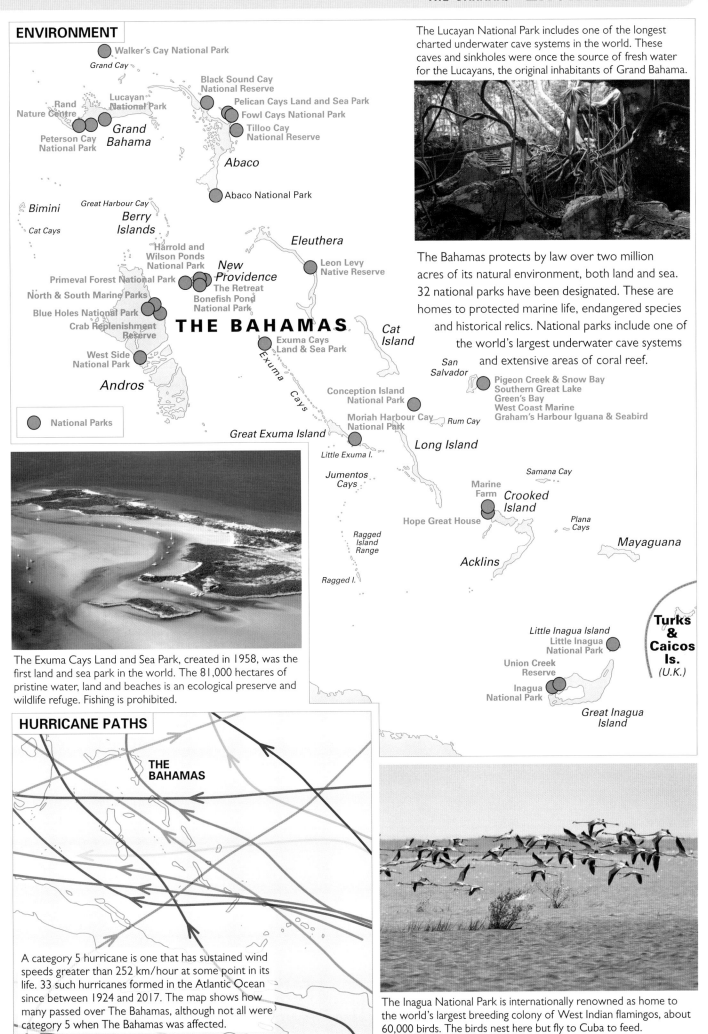

ENVIRONMENT

- Walker's Cay National Park
- Grand Cay
- Black Sound Cay National Reserve
- Lucayan National Park
- Rand Nature Centre
- Pelican Cays Land and Sea Park
- Fowl Cays National Park
- Tilloo Cay National Reserve
- Peterson Cay National Park
- **Grand Bahama**
- *Abaco*
- Abaco National Park
- *Bimini*
- Great Harbour Cay
- *Cat Cays*
- **Berry Islands**
- Harrold and Wilson Ponds National Park
- **New Providence**
- Leon Levy Native Reserve
- *Eleuthera*
- Primeval Forest National Park
- North & South Marine Parks
- The Retreat
- Bonefish Pond National Park
- Blue Holes National Park
- Crab Replenishment Reserve
- **THE BAHAMAS**
- West Side National Park
- *Andros*
- Exuma Cays Land & Sea Park
- *Cat Island*
- *Exuma Cays*
- *San Salvador*
- Pigeon Creek & Snow Bay
- Southern Great Lake
- Green's Bay
- West Coast Marine
- Graham's Harbour Iguana & Seabird
- Conception Island National Park
- Moriah Harbour Cay National Park
- *Rum Cay*
- *Great Exuma Island*
- *Long Island*
- *Little Exuma I.*
- *Jumentos Cays*
- *Samana Cay*
- Marine Farm
- **Crooked Island**
- Hope Great House
- *Plana Cays*
- *Mayaguana*
- *Ragged Island Range*
- *Acklins*
- *Ragged I.*
- National Parks
- Little Inagua Island
- Little Inagua National Park
- **Turks & Caicos Is.** (U.K.)
- Union Creek Reserve
- Inagua National Park
- *Great Inagua Island*

The Lucayan National Park includes one of the longest charted underwater cave systems in the world. These caves and sinkholes were once the source of fresh water for the Lucayans, the original inhabitants of Grand Bahama.

The Bahamas protects by law over two million acres of its natural environment, both land and sea. 32 national parks have been designated. These are homes to protected marine life, endangered species and historical relics. National parks include one of the world's largest underwater cave systems and extensive areas of coral reef.

The Exuma Cays Land and Sea Park, created in 1958, was the first land and sea park in the world. The 81,000 hectares of pristine water, land and beaches is an ecological preserve and wildlife refuge. Fishing is prohibited.

HURRICANE PATHS

THE BAHAMAS

A category 5 hurricane is one that has sustained wind speeds greater than 252 km/hour at some point in its life. 33 such hurricanes formed in the Atlantic Ocean since between 1924 and 2017. The map shows how many passed over The Bahamas, although not all were category 5 when The Bahamas was affected.

The Inagua National Park is internationally renowned as home to the world's largest breeding colony of West Indian flamingos, about 60,000 birds. The birds nest here but fly to Cuba to feed.

ANCESTORS

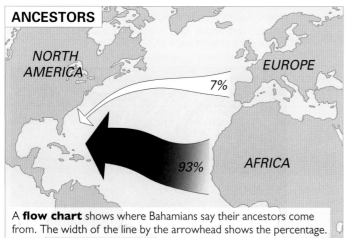

A **flow chart** shows where Bahamians say their ancestors come from. The width of the line by the arrowhead shows the percentage.

GROWTH

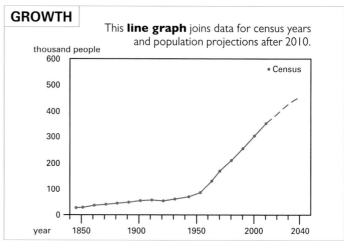

This **line graph** joins data for census years and population projections after 2010.

AGE & GENDER

This type of graph is called a **population pyramid**. Females are shown in blue, males in red.

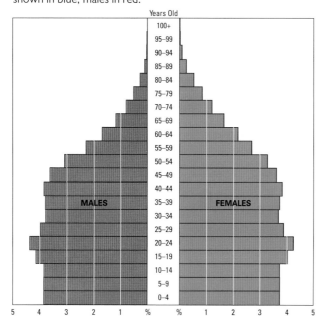

ISLAND DATA

Island	Population (2010 census)	Area (sq km)	Population Density (People/ sq km)
All Bahamas	351,461	13,913	25
New Providence	246,329	207	1,189
Grand Bahama	51,368	1,373	37
Abaco	17,224	1,681	10
Acklins	565	497	1
Andros	7,490	5,957	1
Berry Islands	807	31	26
Bimini	1,988	28	70
Cat Island	1,522	388	4
Crooked Island	330	218	2
Eleuthera, Harbour Island & Spanish Wells	11,515	518	22
Exuma and Cays	6,928	290	24
Inagua	913	1,551	1
Long Island	3,094	596	5
Mayaguana	277	285	1
Ragged Island	72	36	2
San Salvador & Rum Cay	1,039	233	4

DISTRIBUTION

The segments of this **pie chart**, or **divided circle**, are in proportion to the number of Bahamians in each place.

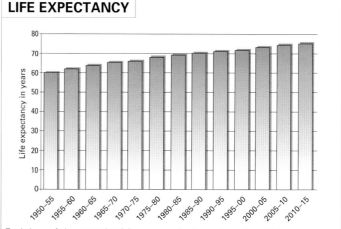

Total population in 2010 census
351,461

LIFE EXPECTANCY

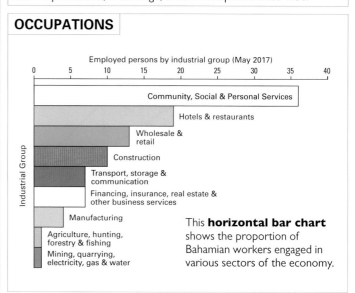

Each bar of this **vertical bar graph** shows how long Bahamians could expect to live, on average, in each time period since 1950.

OCCUPATIONS

Employed persons by industrial group (May 2017)

Community, Social & Personal Services
Hotels & restaurants
Wholesale & retail
Construction
Transport, storage & communication
Financing, insurance, real estate & other business services
Manufacturing
Agriculture, hunting, forestry & fishing
Mining, quarrying, electricity, gas & water

This **horizontal bar chart** shows the proportion of Bahamian workers engaged in various sectors of the economy.

GRAND BAHAMA *1955 agreement*

The Hawksbill Creek Agreement of 1955 granted around 200 sq km of what was then the almost unpopulated island to US investors, together with tax-free status. The result is the city of Freeport, a deep-water harbour, and tourist resort. As a result, the population of Grand Bahama has grown more than 10 times to 50,000.

ABACO *Loyalist settlement 1783*

1,500 people who were loyal to the British Crown left New York in 1783 as the United States fought for its independence. Other Loyalists joined later. They established Hope Town, Marsh Harbour, Cherokee and other towns. The landing is commemorated in a sculpture garden in Green Turtle Cay.

ELEUTHERA *Taíno homeland, more than 1000 years ago*

Taíno peoples migrated from northern parts of South America and settled across the Bahama islands, including Eleuthera. They are known as Lucayan Taíno. Recent research by archaeologists on human remains found in Preacher's Cave on Eleuthera, shows the significance of caves in Lucayan Taíno beliefs and culture, and suggests that some Taíno survived European colonisation. Preacher's Cave served as a haven for shipwrecked Puritans who fled Bermuda. These refugees established the first non-Lucayan settlement in 1648 at Cupid's Cay.

NEW PROVIDENCE
Nassau, capital of the Bahamas since 1718

Charles Town was established in 1670 (named after England's King Charles II) and renamed Nassau in 1695, in honour of the Dutch House of Nassau which gave England its King William II. From 1718, Nassau became the seat of colonial government of The Bahamas and is now the capital city. The buildings in Parliament Square were constructed in 1813 and today contain the House of Assembly and the Senate.

Grand Bahama   *Abaco*

New Providence   *Eleuthera*

Nassau

Andros

San Salvador

THE BAHAMAS

Inagua

ANDROS *For a century, the island with The Bahamas' biggest industry*

For a century from 1840, Andros was the focus for the sponge industry – thousands of spongers migrated to the island from Greece to fish on the Great Bahama Bank to the west of Andros. For decades, Andros sponging was The Bahamas' largest industry. Although sponges were largely wiped out in the 1930s by a Red Tide infestation (and most spongers migrated to Florida) the tradition survives today in Red Bays.

INAGUA *Four centuries as salt producer*

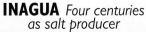

From the days of the Spanish in the 1600s, salt was produced on Great Inagua and shipped to the Spanish colonies. Today the American Morton Salt Company owns the saltworks of over 80 salt ponds. The site comprises 1200 sq km, produces about a half-million kilograms of salt per year, one of the largest in the Americas.

SAN SALVADOR
Columbus's first landfall in the Americas 1492

This map of Guanahani was published in 1685 in Alain Manesson Mallet's Atlas of the World. This island, named by the Lucayan Taínos, is believed to be the first place that Christopher Columbus landed in the Americas. The Spanish renamed it San Salvador. The British called it Watling Island for many decades.

TOURISM FACILITIES & ATTRACTIONS

Tourism is the mainstay of the Bahamian economy, contributing over 55% to the country's total gross domestic product. It is estimated that over US$3 billion is spent annually by almost 1.5 million stopover visitors and nearly 5 million cruise visitors.

The principal tourism islands are New Providence, Grand Bahama, Abaco, Exuma and Eleuthera, although most of the other Family Islands have a share of visitors. 85% of all visitors to The Bahamas originate in North America.

The Bahamas is one of the world's most visited cruise destinations: 30% of all cruise visitors to the Caribbean region visit The Bahamas.

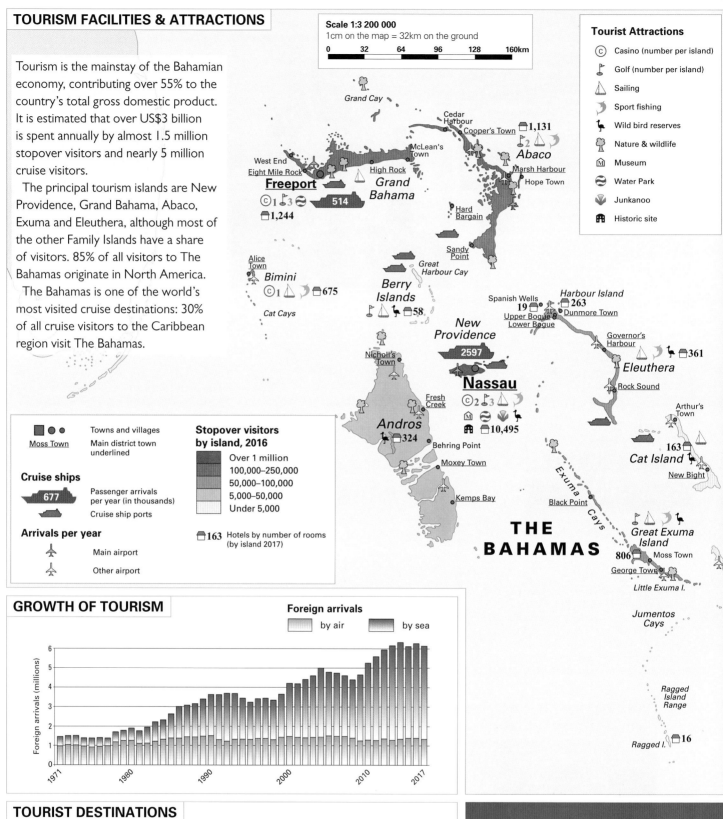

Scale 1:3 200 000
1cm on the map = 32km on the ground

0   32   64   96   128   160km

Tourist Attractions

- Ⓒ Casino (number per island)
- Golf (number per island)
- Sailing
- Sport fishing
- Wild bird reserves
- Nature & wildlife
- Ⓜ Museum
- Water Park
- Junkanoo
- Historic site

Towns and villages

Moss Town — Main district town underlined

Cruise ships

677 — Passenger arrivals per year (in thousands)

— Cruise ship ports

Arrivals per year

- Main airport
- Other airport

Stopover visitors by island, 2016

- Over 1 million
- 100,000–250,000
- 50,000–100,000
- 5,000–50,000
- Under 5,000

163 — Hotels by number of rooms (by island 2017)

GROWTH OF TOURISM

Foreign arrivals
☐ by air   ■ by sea

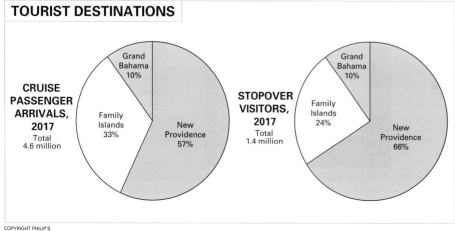

TOURIST DESTINATIONS

CRUISE PASSENGER ARRIVALS, 2017
Total 4.6 million

- Grand Bahama 10%
- Family Islands 33%
- New Providence 57%

STOPOVER VISITORS, 2017
Total 1.4 million

- Grand Bahama 10%
- Family Islands 24%
- New Providence 66%

The Lynden Pindling International Airport in New Providence is the largest and busiest in The Bahamas. Around 80,000 aircraft and 6 million passengers pass through every year.

PROFILE OF A TYPICAL VISITOR TO THE BAHAMAS

- For 74%, the main purpose was to vacation.
- 85% selected The Bahamas to enjoy its beaches.
- 86% intend to repeat visit in one to five years.
- 52% aged between 22 and 54 years.
- For 56%, the annual income of their household exceeds $75,000 per year.
- 45% travel to The Bahamas as a party of two.

The Atlantis Resort on Paradise Island covers half the island. It includes a huge marine park, dolphin lake, mega-yacht marina, multiple aquariums, a casino and a convention centre. The casino is one of the Caribbean's largest. 85 table games and 700 slot machines are ready for visiting gamblers 24 hours a day, every day.

Junkanoo parades are one of the most visible expressions of Bahamian culture. Downtown Nassau's Junkanoo Museum gives a chance to explore this Bahamian tradition even when missing the Christmas and Summer Parades.

ORIGIN OF STOPOVER TOURISTS

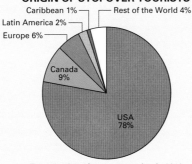

Caribbean 1%
Latin America 2%
Europe 6%
Rest of the World 4%
Canada 9%
USA 78%

Total number of stopover tourists in 2016
1,500,000

San Salvador
Cockburn Town
250

Rum Cay   0   Rum Cay

Family Islands
1521

192
Long Island
Deadman's Cay

Clarence Town   Crooked Island   Samana Cay

34
Moss Town
Colonel Hill

Plana Cays

Albert Town   Mason Bay

Acklins   56

Mayaguana
13
Abraham's Bay

Little Inagua Island

35

Matthew Town   Great Inagua Island

Windsurfers are competing in Montagu Bay in Nassau. Yacht racing is an extremely popular sport in The Bahamas. Native dinghies and sloops compete in annual regattas held on many of the islands.

Six golf courses in The Bahamas have been rated in the top 50 in the region. They attract golfing world stars such as Tiger Woods.

Fishing caters to first-timers casting a line, to deep-sea enthusiasts eager to beat world record catches. This barracuda was caught off Long Island.

The Commonwealth of The Bahamas occupies a total continental shelf area of approximately 116,550 sq km. It comprises 3,000 small islands and cays with a land area of about 13,935 sq km. These islands and cays are spread over an area of some 230,000 sq km and are located on 14 plateaus separated from each other and from Florida, Cuba and Hispaniola by depths of 360–3,600 metres.

Agriculture and fisheries account for 1.6% of the Gross Domestic Product of The Bahamas and employ over 3% of the working population. Of the 1.6% about 1.2% is fishing. Only about 1% of the land area of the country is cultivated because the poor soils and rainfall limit farming.

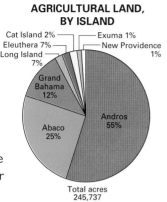

AGRICULTURAL LAND, BY ISLAND

Cat Island 2%
Eleuthera 7%
Long Island 7%
Grand Bahama 12%
Abaco 25%
Andros 55%
Exuma 1%
New Providence 1%

Total acres 245,737

CROP & LIVESTOCK PRODUCTION

Crop and livestock production, while small, increased significantly since 2000

Production index (100 = 2004–2006 data)

Crop production index
Livestock production index

AGRICULTURE & FISHING AREAS

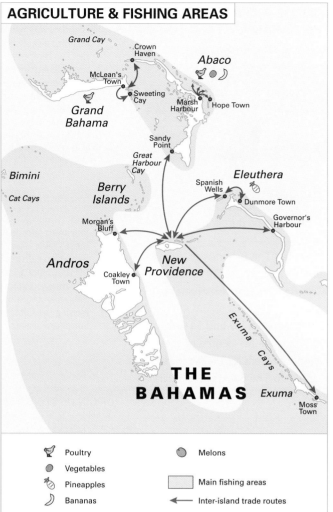

Grand Cay
Crown Haven
McLean's Town
Sweeting Cay
Abaco
Marsh Harbour
Hope Town
Grand Bahama
Sandy Point
Great Harbour Cay
Bimini
Berry Islands
Spanish Wells
Eleuthera
Dunmore Town
Governor's Harbour
Cat Cays
Morgan's Bluff
Andros
New Providence
Coakley Town
THE BAHAMAS
Exuma Cays
Exuma
Moss Town

Poultry
Vegetables
Pineapples
Bananas
Melons
Main fishing areas
Inter-island trade routes

Agriculture for The Bahamas, if practised, is often a part-time occupation and the crops are grown for personal use. Some fish and vegetables are exported. Fish and crustaceans account for over 90% of agri-food exports and are destined for the EU, USA and Canada.

CRUSTACEAN & FISH PRODUCTION

	Name	Weight (tonnes)	Value (US$)
	Spiny lobster	6,977	70,366,282
	Snapper	569	2,848,370
	Conch	379	3,051,282
	Grouper	223	2,058,415
	Jack	84	619,452
	Grunt	39	102,967
	Stone crab	31	582,527
	Turtle (Loggerhead)	1	3,880
	Other	29	114,235
	TOTAL (2007 data)	8,331	79,747,410

The Caribbean Spiny Lobster (or Crawfish) is an important commercial species in Bahamian waters. Over five million lobster tails are exported every year.

Coat of Arms

The flag of The Commonwealth of The Bahamas was raised for the first time on the day when independence was declared. Gold represents the sun, acquamarine blue the sea, and black the strength of Bahamian people.

The Bahamas adopted one each of its trees, flowers, birds and fish to be four national symbols. Some of these symbols are included in the national coat of arms. Which colours of the flag and which symbols are not represented in the national coat of arms?

FLAG, COAT OF ARMS, SYMBOLS

On July 10, 1973, thousands of Bahamians sang the National Anthem for the first time: *Lift up your head to the rising sun Bahamaland*. This was after Prince Charles, representing the United Kingdom, handed the instruments of independence to the first Prime Minister of The Bahamas, Lynden Pindling.

Yellow elder,
national flower

Blue marlin,
national fish

Pink flamingoes,
national bird

Lignam vitea,
national tree

RELIGION

No religion
10%

Other faiths* 4%

Methodist 4%

Seventh-day Adventist
5%

Church of God
5%

Penticostal
8%

Baptist
35%

Roman
Catholic
14%

Anglican
15%

* Smaller percentages of Bahamians profess
Greek Orthodox, Jewish, Baha'i, Jehovah's Witness,
Muslim, Obeah, Rastafarian and Hindu religions

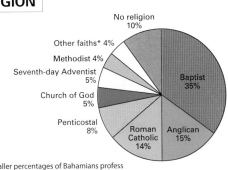

Funeral service of leading Harbour Island community member, Ma Ruby, a member of the St John's Anglican Church. The church was built in 1768.

JUNKANOO

Junkanoo is a street parade with music, dance and costumes that originated in the Akan culture of West Africa. It was remembered and recreated in The Bahamas by enslaved Africans. Parades happen in many islands across The Bahamas every Boxing Day and New Year's Day. The biggest Junkanoo parade is in Nassau.

ORIGIN

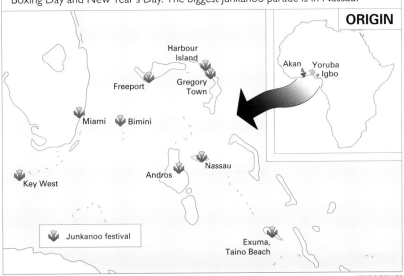

Harbour
Island

Akan   Yoruba
Igbo

Freeport   Gregory
Town

Miami   Bimini

Key West   Andros   Nassau

Exuma,
Taino Beach

Junkanoo festival

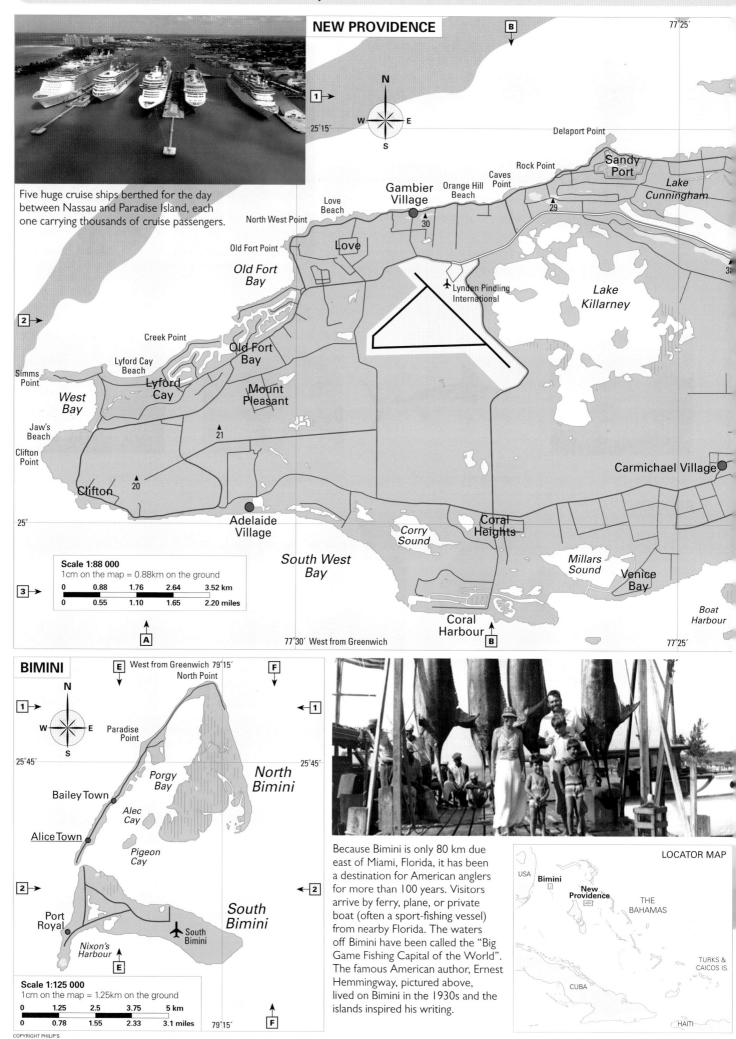

Five huge cruise ships berthed for the day between Nassau and Paradise Island, each one carrying thousands of cruise passengers.

NEW PROVIDENCE

77°25′

25°15′

Delaport Point

Love Beach

Gambier Village

Orange Hill Beach

Caves Point

Rock Point

Sandy Port

Lake Cunningham

North West Point

Love

Old Fort Point

Old Fort Bay

Lynden Pindling International

Lake Killarney

Creek Point

Old Fort Bay

Lyford Cay Beach

Simms Point

Lyford Cay

Mount Pleasant

West Bay

Jaw's Beach

21

Clifton Point

20

Clifton

Carmichael Village

25°

Adelaide Village

Corry Sound

Coral Heights

Millars Sound

Venice Bay

South West Bay

Coral Harbour

Boat Harbour

Scale 1:88 000
1cm on the map = 0.88km on the ground

| 0 | 0.88 | 1.76 | 2.64 | 3.52 km |
| 0 | 0.55 | 1.10 | 1.65 | 2.20 miles |

77°30′ West from Greenwich

77°25′

BIMINI

West from Greenwich 79°15′

North Point

25°45′

Paradise Point

25°45′

Porgy Bay

Bailey Town

North Bimini

Alec Cay

Alice Town

Pigeon Cay

Port Royal

South Bimini

South Bimini

Nixon's Harbour

Scale 1:125 000
1cm on the map = 1.25km on the ground

| 0 | 1.25 | 2.5 | 3.75 | 5 km |
| 0 | 0.78 | 1.55 | 2.33 | 3.1 miles |

79°15′

Because Bimini is only 80 km due east of Miami, Florida, it has been a destination for American anglers for more than 100 years. Visitors arrive by ferry, plane, or private boat (often a sport-fishing vessel) from nearby Florida. The waters off Bimini have been called the "Big Game Fishing Capital of the World". The famous American author, Ernest Hemmingway, pictured above, lived on Bimini in the 1930s and the islands inspired his writing.

LOCATOR MAP

USA

Bimini

New Providence

THE BAHAMAS

CUBA

TURKS & CAICOS IS.

HAITI

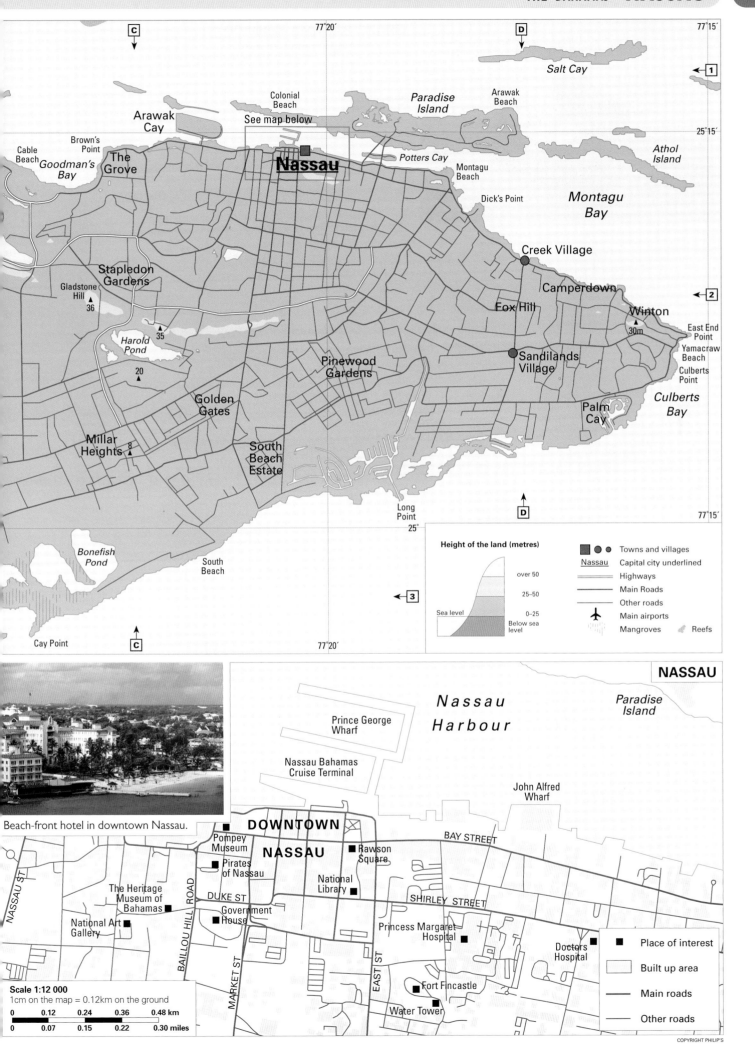

77°20′   D   77°15′

Salt Cay

← 1

Arawak Cay

Brown's Point

Cable Beach

Goodman's Bay

The Grove

Colonial Beach

See map below

Nassau

Paradise Island

Arawak Beach

25°15′

Potters Cay

Montagu Beach

Athol Island

Dick's Point

Montagu Bay

Stapledon Gardens

Gladstone Hill ▲ 36

Harold Pond

▲ 35

▲ 20

Creek Village

Camperdown

Fox Hill

Winton ▲ 30m

East End Point

Yamacraw Beach

Culberts Point

Pinewood Gardens

Sandilands Village

Golden Gates

Millar Heights ▲ 8

South Beach Estate

Palm Cay

Culberts Bay

Long Point

D

77°15′

Bonefish Pond

South Beach

25°

← 3

Height of the land (metres)

over 50

25–50

Sea level   0–25

Below sea level

■ ● ● Towns and villages

<u>Nassau</u> Capital city underlined

Highways

Main Roads

Other roads

✈ Main airports

Mangroves   Reefs

Cay Point   C   77°20′

NASSAU

Nassau Harbour

Paradise Island

Prince George Wharf

Nassau Bahamas Cruise Terminal

John Alfred Wharf

DOWNTOWN

NASSAU

Pompey Museum

Pirates of Nassau

The Heritage Museum of Bahamas

National Art Gallery

Rawson Square

BAY STREET

National Library

DUKE ST

Government House

SHIRLEY STREET

Princess Margaret Hospital

Doctors Hospital

■ Place of interest

Built up area

Main roads

Other roads

Fort Fincastle

Water Tower

NASSAU ST   BAILLOU HILL ROAD   MARKET ST   EAST ST

Beach-front hotel in downtown Nassau.

Scale 1:12 000
1cm on the map = 0.12km on the ground

0   0.12   0.24   0.36   0.48 km

0   0.07   0.15   0.22   0.30 miles

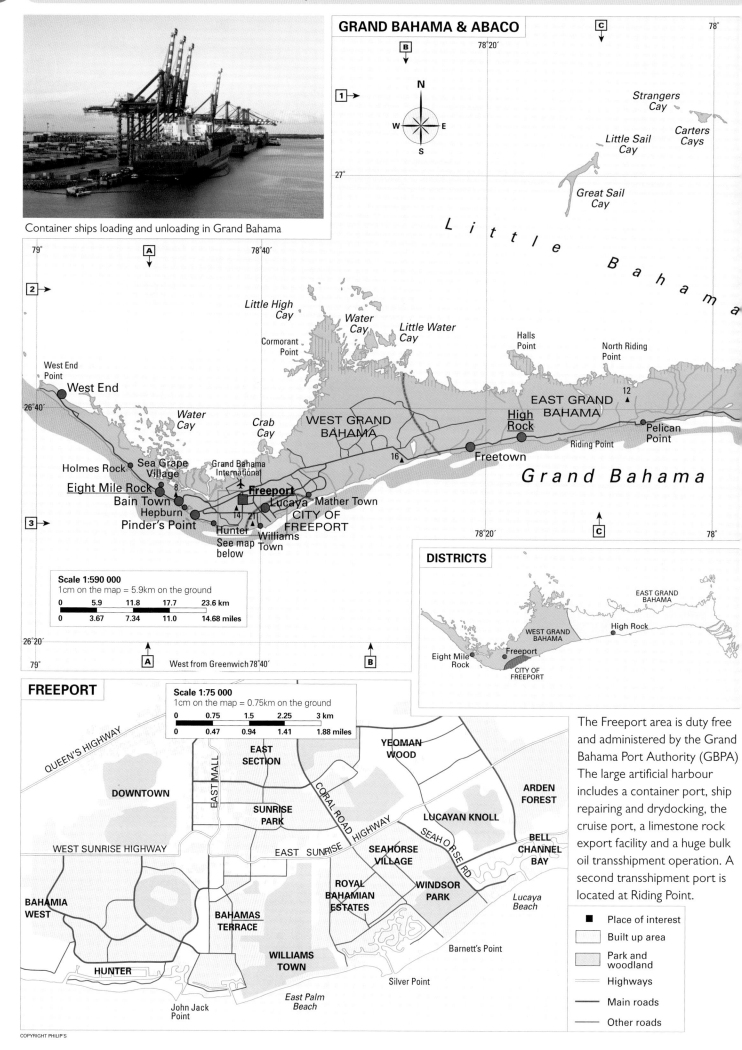

Container ships loading and unloading in Grand Bahama

GRAND BAHAMA & ABACO

N
W · E
S

Strangers Cay

Little Sail Cay

Carters Cays

Great Sail Cay

L i t t l e   B a h a m a

78°20′

78°

27°

West End Point

West End

Little High Cay

Water Cay

Little Water Cay

Halls Point

North Riding Point

Cormorant Point

EAST GRAND BAHAMA

12

26°40′

Water Cay

Crab Cay

WEST GRAND BAHAMA

High Rock

Pelican Point

Riding Point

Holmes Rock

Sea Grape Village

Grand Bahama International

16

Freetown

G r a n d   B a h a m a

Eight Mile Rock

8

Freeport

Bain Town

Lucaya

Mather Town

Hepburn

14

21

CITY OF FREEPORT

Pinder's Point

Hunter

Williams Town

See map below

78°20′

78°

Scale 1:590 000
1cm on the map = 5.9km on the ground

0	5.9	11.8	17.7	23.6 km
0	3.67	7.34	11.0	14.68 miles

26°20′

79°   West from Greenwich 78°40′

DISTRICTS

EAST GRAND BAHAMA

WEST GRAND BAHAMA

Eight Mile Rock

Freeport

High Rock

CITY OF FREEPORT

FREEPORT

Scale 1:75 000
1cm on the map = 0.75km on the ground

0	0.75	1.5	2.25	3 km
0	0.47	0.94	1.41	1.88 miles

QUEEN'S HIGHWAY

EAST SECTION

EAST MALL

YEOMAN WOOD

DOWNTOWN

CORAL ROAD HIGHWAY

ARDEN FOREST

SUNRISE PARK

LUCAYAN KNOLL

WEST SUNRISE HIGHWAY

EAST SUNRISE

SEAHORSE RD

SEAHORSE VILLAGE

BELL CHANNEL BAY

ROYAL BAHAMIAN ESTATES

WINDSOR PARK

Lucaya Beach

BAHAMIA WEST

BAHAMAS TERRACE

Barnett's Point

HUNTER

WILLIAMS TOWN

Silver Point

East Palm Beach

John Jack Point

The Freeport area is duty free and administered by the Grand Bahama Port Authority (GBPA) The large artificial harbour includes a container port, ship repairing and drydocking, the cruise port, a limestone rock export facility and a huge bulk oil transshipment operation. A second transshipment port is located at Riding Point.

■ Place of interest
☐ Built up area
▨ Park and woodland
═ Highways
━ Main roads
─ Other roads

Height of the land (metres)

over 50

25–50

0–25

Sea level

Below sea level

■ ● ● Towns and villages

<u>Freeport</u> Main town underlined

Main roads

Other roads

✈ Main airports

District boundaries

Mangroves ⌁ Reefs

LOCATOR MAP

USA

Grand Bahama

Abaco

THE BAHAMAS

TURKS & CAICOS IS.

CUBA

HAITI

Paw Paw Cays

Fish Cays

Moraine Cay

Allans-Pensacola Cay

Spanish Cay

Crown Haven

Fox Town

Mount Hope

Cedar Harbour

Wood Cay

NORTH ABACO

Little Abaco Island

Cross Cay's

Big Jerry Cay

B a n k

August Cay

Big Harbour Cay

McLeans Town

Little Harbour Cay

Sweetings Cay

Deep Water Cay

Sweetings Town

Lighbourne Cay

Michael's Cay

Long Cay

P r o v i d e n c e   C h a n n e l

Powell Cay

Cooper's Town

Fire Road Village

Manjack Cay

Randells Cay

Rocky Harbour Cays

Blackwood Village

H O P E   T O W N

A T L A N T I C   O C E A N

Green Turtle Cay

New Plymouth

S e a   o f

Treasure Cay ✈

Whale Cay

Great Guana Cay

Scotland Cay

Man-O-War Cay

Treasure Cay

A b a c o

Joe Downer Cays

A b a c o

30 29

Dundas Town

Hope Town Point

Central Pines Estate

Leonard M Thompson International ✈

North End

Marsh Harbour

Hope Town

Elbow Cay

Lubbers Quarters

The Marls

Spring City

CENTRAL ABACO

Channel Cay

Tilloo Cay

37

Lynyard Cay

Hard Bargain

Moore's Island ✈

The Bight

Moore's Island

Wood Cay

Lake City

Little Harbour

Cherokee

Great Abaco Island

Duck Cay

Eight Mile Bay

Cornwall Point

Castaway Cay (Private) ✈

Gorda Cay (Castaway Island)

Crossing Rocks

Sandy Point

SOUTH ABACO

Sandy Point ✈

Hole in the Wall

DISTRICTS

Cooper's Town

NORTH ABACO

HOPE TOWN

Marsh Harbour

Hope Town

Hard Bargain

MOORE'S ISLAND

CENTRAL ABACO

Sandy Point

SOUTH ABACO

Hope Town Harbour, a haven for sailing boats, with its iconic lighthouse top left.

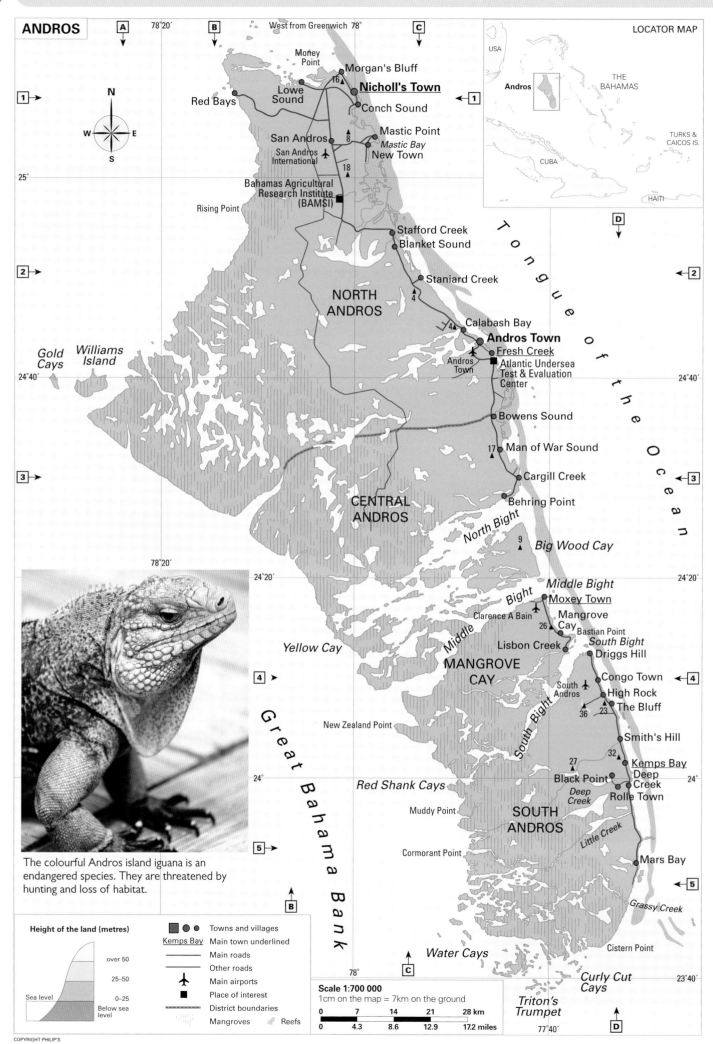

ANDROS

LOCATOR MAP

West from Greenwich

USA
Andros
THE BAHAMAS
TURKS & CAICOS IS.
CUBA
HAITI

Money Point
Morgan's Bluff
Nicholl's Town
Lowe Sound
Conch Sound
Red Bays
Mastic Point
Mastic Bay
San Andros
New Town
San Andros International
Bahamas Agricultural Research Institute (BAMSI)
Rising Point
Stafford Creek
Blanket Sound
Staniard Creek
NORTH ANDROS
Calabash Bay
Andros Town
Fresh Creek
Andros Town
Atlantic Undersea Test & Evaluation Center
Gold Cays
Williams Island
Bowens Sound
Man of War Sound
Cargill Creek
Behring Point
CENTRAL ANDROS
North Bight
Big Wood Cay
Middle Bight
Bight
Moxey Town
Clarence A Bain
Mangrove Cay
Bastian Point
Lisbon Creek
South Bight
Driggs Hill
MANGROVE CAY
South Andros
Congo Town
High Rock
The Bluff
Smith's Hill
Kemps Bay
Black Point
Deep Creek
Deep Creek
Rolle Town
SOUTH ANDROS
Little Creek
Mars Bay
Grassy Creek
Cistern Point

Tongue of the Ocean

Great Bahama Bank

South Bight

Yellow Cay
New Zealand Point
Red Shank Cays
Muddy Point
Cormorant Point
Water Cays
Triton's Trumpet
Curly Cut Cays

The colourful Andros island iguana is an endangered species. They are threatened by hunting and loss of habitat.

Height of the land (metres)
over 50
25–50
Sea level
0–25
Below sea level

Towns and villages
Kemps Bay Main town underlined
Main roads
Other roads
Main airports
Place of interest
District boundaries
Mangroves   Reefs

Scale 1:700 000
1cm on the map = 7km on the ground
0   7   14   21   28 km
0   4.3   8.6   12.9   17.2 miles

COPYRIGHT PHILIP'S

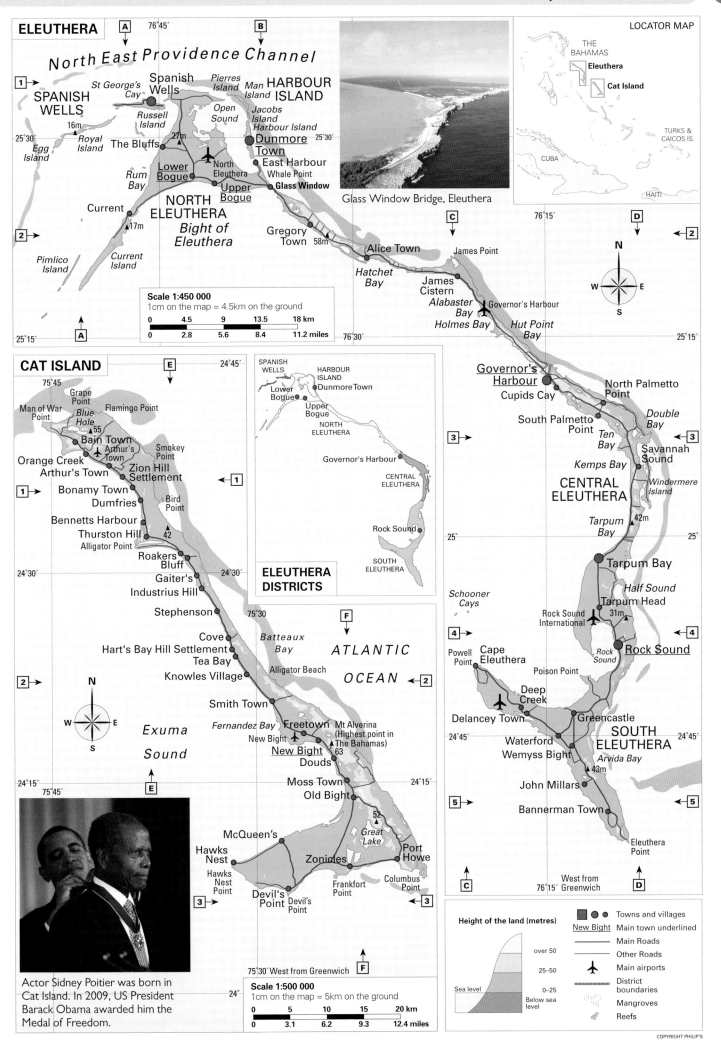

ELEUTHERA

A 76°45′ B

North East Providence Channel

SPANISH WELLS

St George's Cay
Spanish Wells
Pierres Island
Man Island
HARBOUR ISLAND

Russell Island
Open Sound
Jacobs Island
Harbour Island

16m
Royal Island
The Bluffs
27m
North Eleuthera
Dunmore Town

Egg Island
25°30′

Lower Bogue
East Harbour
Whale Point
Glass Window

Rum Bay
NORTH ELEUTHERA
Upper Bogue

Current
Bight of Eleuthera
17m

Pimlico Island
Current Island

Gregory Town
58m

Alice Town
James Point

Hatchet Bay

James Cistern

Alabaster Bay
Governor's Harbour
Holmes Bay
Hut Point Bay

Scale 1:450 000
1cm on the map = 4.5km on the ground

| 0 | 4.5 | 9 | 13.5 | 18 km |

| 0 | 2.8 | 5.6 | 8.4 | 11.2 miles |

76°30′

Glass Window Bridge, Eleuthera

LOCATOR MAP

THE BAHAMAS
Eleuthera
Cat Island

TURKS & CAICOS IS.

CUBA

HAITI

76°15′ D

N
W E
S

Governor's Harbour
Cupids Cay
North Palmetto Point

South Palmetto Point
Double Bay

Ten Bay
Savannah Sound

Kemps Bay
Windermere Island

CENTRAL ELEUTHERA

Tarpum Bay
42m

Tarpum Bay

Half Sound
Tarpum Head
31m

Rock Sound International
Rock Sound

Schooner Cays

Powell Point
Cape Eleuthera
Poison Point

Deep Creek
Greencastle

Delancey Town
SOUTH ELEUTHERA
24°45′

Waterford
Wemyss Bight
Arvida Bay
43m

John Millars

Bannerman Town

Eleuthera Point

C
West from 76°15′ Greenwich
D

CAT ISLAND

E 24°45′

75°45′

Man of War Point
Grape Point
Flamingo Point
Blue Hole
55
Bain Town
Arthur's Town
Smokey Point

Orange Creek
Arthur's Town
Zion Hill Settlement

Bonamy Town
Bird Point

Dumfries
Bennetts Harbour
Thurston Hill
42
Alligator Point

Roakers Bluff
Gaiter's
24°30′
Industrius Hill

Stephenson
75°30′

Cove
Batteaux Bay
F

Hart's Bay Hill Settlement
Tea Bay
ATLANTIC

Knowles Village
Alligator Beach
OCEAN

Smith Town

Exuma
Fernandez Bay
Freetown
Mt Alverina
(Highest point in The Bahamas)
63

New Bight

Sound
New Bight
Douds

Moss Town
24°15′
Old Bight

52
Great Lake

McQueen's
Hawks Nest
Zonicles
Port Howe

Hawks Nest Point
Frankfort Point
Columbus Point

Devil's Point
Devil's Point

E 75°45′ 24°15′

N
W E
S

ELEUTHERA DISTRICTS

SPANISH WELLS
HARBOUR ISLAND
Lower Bogue
Dunmore Town
Upper Bogue
NORTH ELEUTHERA

Governor's Harbour

CENTRAL ELEUTHERA

Rock Sound

SOUTH ELEUTHERA

Actor Sidney Poitier was born in Cat Island. In 2009, US President Barack Obama awarded him the Medal of Freedom.

Scale 1:500 000
1cm on the map = 5km on the ground

| 0 | 5 | 10 | 15 | 20 km |

| 0 | 3.1 | 6.2 | 9.3 | 12.4 miles |

75°30′ West from Greenwich F
24°

Height of the land (metres)

over 50
25–50
Sea level
0–25
Below sea level

Towns and villages
New Bight Main town underlined
Main Roads
Other Roads
Main airports
District boundaries
Mangroves
Reefs

COPYRIGHT PHILIP'S

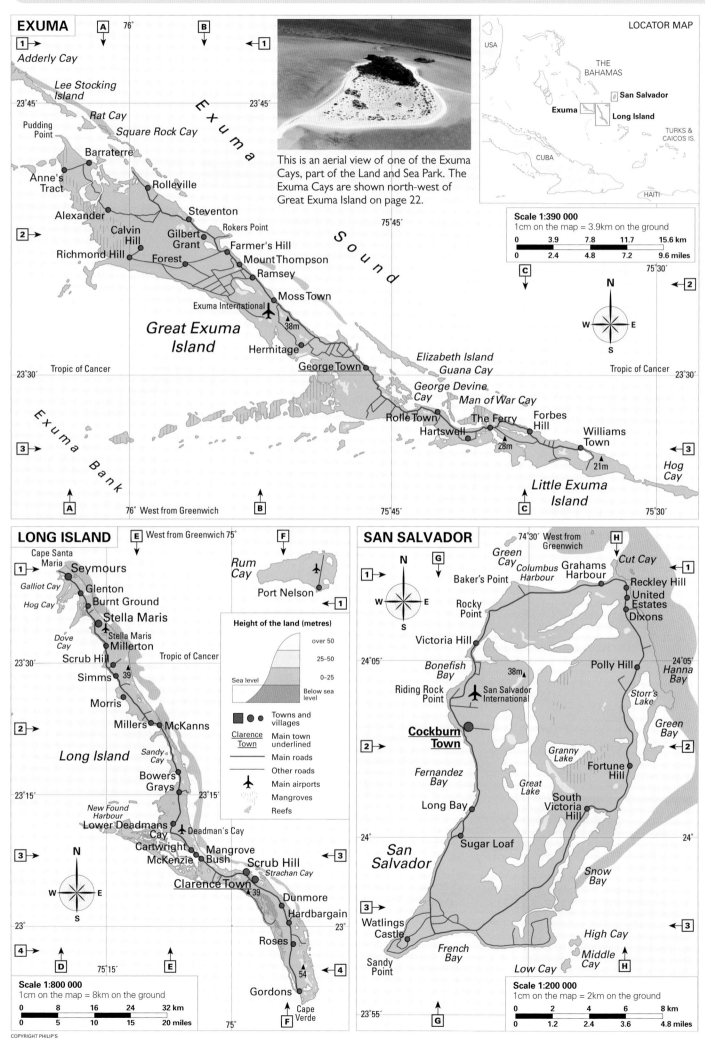

EXUMA

Adderly Cay
Lee Stocking Island
Rat Cay
Square Rock Cay
Pudding Point
Barraterre
Anne's Tract
Rolleville
Alexander
Calvin Hill
Gilbert Grant
Richmond Hill
Forest
Steventon
Rokers Point
Farmer's Hill
Mount Thompson
Ramsey
Moss Town
Exuma International
38m
Hermitage

Great Exuma Island

Tropic of Cancer

Elizabeth Island
Guana Cay
George Town
George Devine Cay
Man of War Cay
Rolle Town
Forbes Hill
The Ferry
Hartswell
28m
Williams Town
21m

Exuma Bank
Exuma Sound

Hog Cay
Little Exuma Island

76° West from Greenwich
75°45′
75°30′

23°45′
23°30′
Tropic of Cancer

LOCATOR MAP

USA
THE BAHAMAS
San Salvador
Exuma
Long Island
TURKS & CAICOS IS.
CUBA
HAITI

This is an aerial view of one of the Exuma Cays, part of the Land and Sea Park. The Exuma Cays are shown north-west of Great Exuma Island on page 22.

Scale 1:390 000
1cm on the map = 3.9km on the ground

| 0 | 3.9 | 7.8 | 11.7 | 15.6 km |
| 0 | 2.4 | 4.8 | 7.2 | 9.6 miles |

N
W E
S

LONG ISLAND

Cape Santa Maria
Seymours
Galliot Cay
Glenton
Hog Cay
Burnt Ground
Stella Maris
Dove Cay
Stella Maris
Millerton
Scrub Hill
Tropic of Cancer
Simms
39
Morris
Millers
McKanns

Long Island

Sandy Cay
Bowers
Grays
New Found Harbour
Lower Deadmans Cay
Deadman's Cay
Cartwright
Mangrove Bush
McKenzie
Scrub Hill
Strachan Cay
Clarence Town
39
Dunmore
Hardbargain
Roses

N
W E
S

Gordons
Cape Verde

West from Greenwich 75°
Rum Cay
Port Nelson

23°30′
23°15′
23°

75°15′
75°

Height of the land (metres)

over 50
25–50
Sea level
0–25
Below sea level

Towns and villages
Clarence Town Main town underlined
Main roads
Other roads
Main airports
Mangroves
Reefs

Scale 1:800 000
1cm on the map = 8km on the ground

| 0 | 8 | 16 | 24 | 32 km |
| 0 | 5 | 10 | 15 | 20 miles |

SAN SALVADOR

N
W E
S

74°30′ West from Greenwich
Green Cay
Columbus Harbour
Grahams Harbour
Cut Cay
Baker's Point
Reckley Hill
United Estates
Rocky Point
Dixons
Victoria Hill
Bonefish Bay
38m
Polly Hill
Hanna Bay
Riding Rock Point
San Salvador International
Storr's Lake
Green Bay
Cockburn Town
Granny Lake
Fortune Hill
Fernandez Bay
Great Lake
South Victoria Hill
Long Bay
Sugar Loaf

San Salvador

Snow Bay
Watlings Castle
High Cay
Sandy Point
French Bay
Low Cay
Middle Cay

24°05′
24°
23°55′

Scale 1:200 000
1cm on the map = 2km on the ground

| 0 | 2 | 4 | 6 | 8 km |
| 0 | 1.2 | 2.4 | 3.6 | 4.8 miles |

CROOKED ISLAND & ACKLINS

74° West from Greenwich C

74°20' A

Landrail Point

Pitts Town
Seaview
Moss Town
Fairfield
Cabbage Hill
Colonel Hill
Colonel Hill
Church Grove
Colonel Hill
Majors Cay
Majors
Browns
Bullet Hill

Crooked Island

▲15m

1

1

22°40'
Goat Cay
Rat Cay
Lucian Cay

Lovely Bay
Chesters

Lovely Bay

22°40'

Albert Town
Long Cay

The Bight of Acklins

Pinefield
Anderson
Hardhill

North Cay

Fish Cay

Snug Corner
Mason's Bay
Goodwill
Mason's Bay
Creek Point

N

Spring Point
Spring Point

Florida
Abrahams Bay

Delectable Bay
43 ▲

Jamaica Cay

W · E

S

Acklins

3

22°20'

22°20'

Binnacle Hill

Rokers Cay

Salina Point

Southwest Point

74°
Bird Rock Lighthouse, Crooked Island

C

Mira Por Vos Passage

Castle Island

A
74°20'
B

MAYAGUANA

73° E 72°50' F

73°10' D

ATLANTIC OCEAN

N

22°30'
Northwest Point

Pirates Well

Mayaguana

22°30'

1

Betsy Bay

W · E

1

Devils Point

Deans Bay

Abraham's Bay

40 ▲

Northeast Point

22°20'

Horse Pond Bay

22°20'

2

2

73°10' D 73° E West 72°50' from Greenwich

Height of the land (metres)

over 50

25–50

Sea level

0–25

Below sea level

◼ ● ● Towns and villages
Nassau Capital city underlined
── Main roads
── Other roads
── National Park boundary
✈ Main airports
〰 Mangroves ⚓ Reefs

Scale 1:700 000
1cm on the map = 7km on the ground

0 · 7 · 14 · 21 · 28 km
0 · 4.3 · 8.6 · 12.9 · 17.2 miles

RAGGED ISLANDS

75°40' J

75°50' G

Jumentos Cay

1

Water Cay

23°

1

Torzon Cay

N

2

2

W · E

2

S

22°50'

Flamingo Cay

Man of War Cay

3

3

Jamaica Cay

West from Greenwich
75°50' 75°40'

Ragged

Island Range

22°30'

South Channel Cay
Nurse Cay

Bueno Vista Cay

5

5

Raccoon Cay

22°20'

Johnson Cay

Double Breasted Cay
Margaret Cay
Maycock Cay

Hog Cay

6

Ragged Island

6

Duncan Town
Duncan Town ✈

Little Ragged Island

22°

G
75°50'
H
75°40'
J

INAGUA

L 73°00'

73°20'

Little Inagua

1

29 ▲

30 ▲

N

1

W · E

S

21°20'

Northeast Point

21°20'

73°40' K

Great Inagua

Palacca Point

Mutton Fish Point

Cool Bay

Ocean Bight

2

Sheep Cay

Inagua National Park

2

Northwest Point

Man of War Bay

14 ▲

21°00'

Lake Rosa (Lake Windsor)

33 ▲

21°00'

✈ Inagua

31 ▲

Lantern Head Harbour

Matthew Town

South Bay

Southwest Point

Conch Shell Point

3

73°40' K

73°20' West from Greenwich L

73°00'

LOCATOR MAP

USA

THE BAHAMAS

Ragged Islands

Crooked I.

Mayaguana

Acklins I.

TURKS & CAICOS IS.

CUBA

Inagua

HAITI

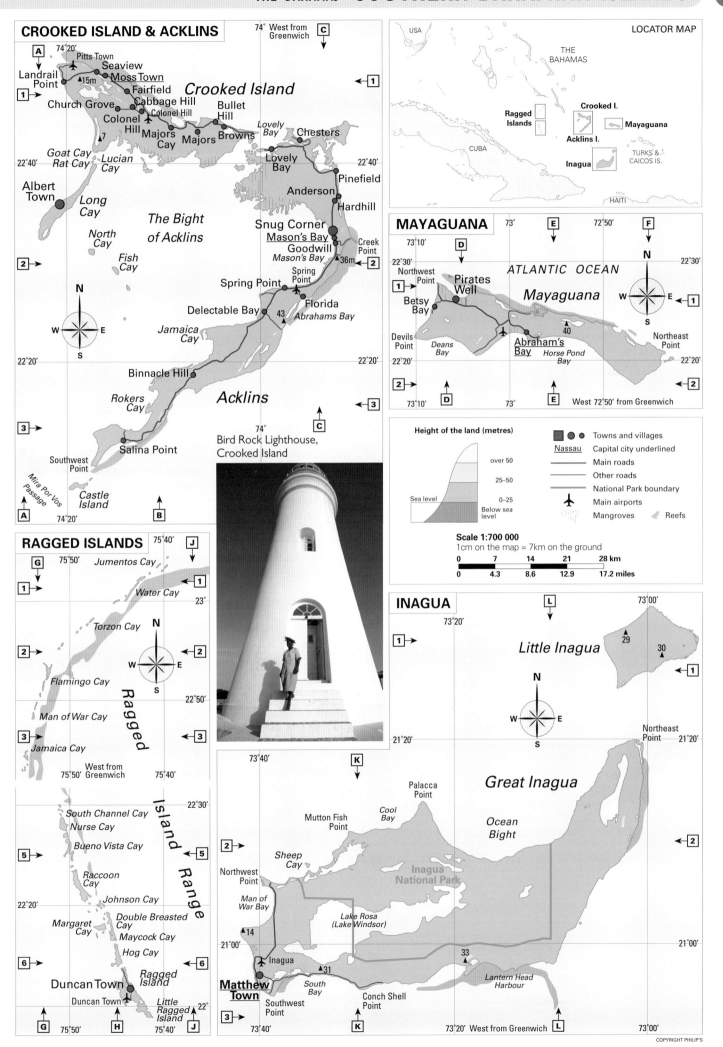

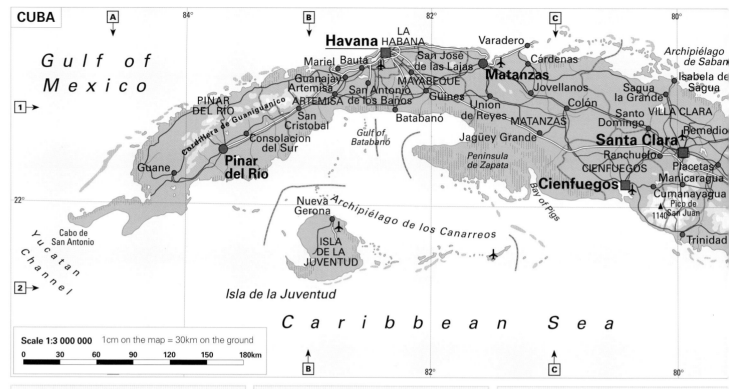

CUBA

Independence 1902
Capital: Havana
Area: 110,860 sq km
Population: 11,239,000 (est. 2018)
Languages: Spanish
Sources of national income: tourism, nickel, colbalt, tobacco, fish, petroleum, citrus, coffee, sugar

HAITI

Independence 1804
Capital: Port-au-Prince
Area: 27,275 sq km
Population: 10,912,000 (est. 2018)
Languages: French (official), Creole (official)
Sources of national income: tourism, clothing, coffee, oils, cocoa

DOMINICAN REPUBLIC

Independence 1844
Capital: Santo Domingo
Area: 48,730 sq km
Population: 10,169,000 (est. 2018)
Languages: Spanish
Sources of national income: tourism, sugar, gold, silver, coffee, cocoa, tobacco, meats, consumer goods

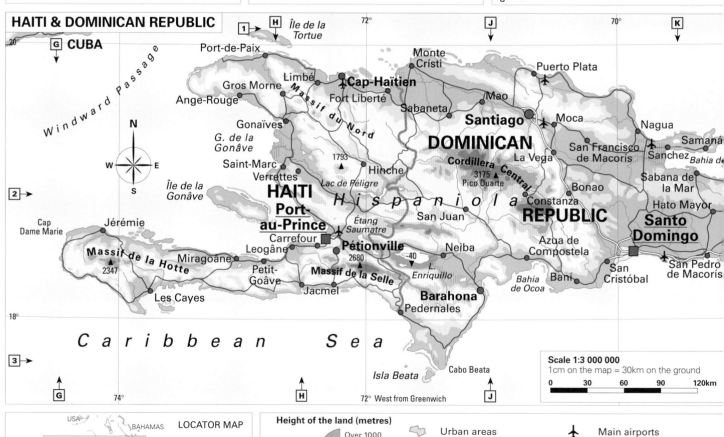

LOCATOR MAP

Height of the land (metres)

Over 1000	
400–1000	
200–400	
100–200	
0–100	
Sea level	
Below sea level	

Urban areas
Towns and villages
Havana Capital city underlined
Highways
Roads
Railways

Main airports
International boundaries
Administrative boundaries
Mangroves
Reefs

PUERTO RICO (U.S.A.)

U.S.A. Unincorporated Territory
Capital: San Juan
Area: 9,104 sq km
Population: 3,678,000 (est.2018)
Languages: Spanish, English
Sources of national income: sugar cane,
coffee, fruit, food products, chemicals, electronics, clothing, tourism

UNITED STATES VIRGIN ISLANDS

U.S.A. Unincorporated Territory
Capital: Charlotte Amalie
Area: 352 sq km
Population: 106,000 (est. 2018)
Languages: English (official), Spanish, Creole
Sources of national income: tourism

PUERTO RICO & US VIRGIN ISLANDS

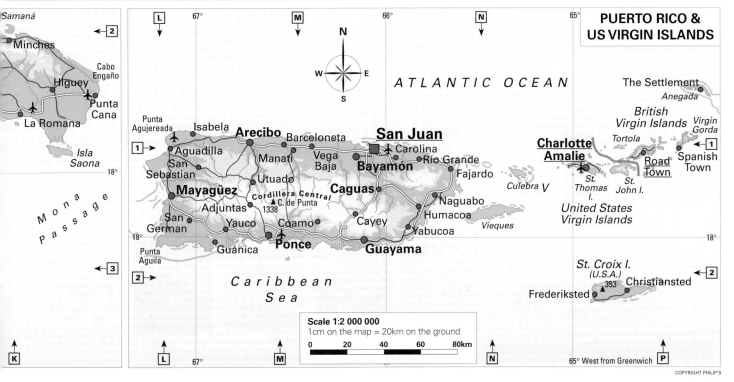

Scale 1:2 000 000
1cm on the map = 20km on the ground

0   20   40   60   80km

COPYRIGHT PHILIP'S

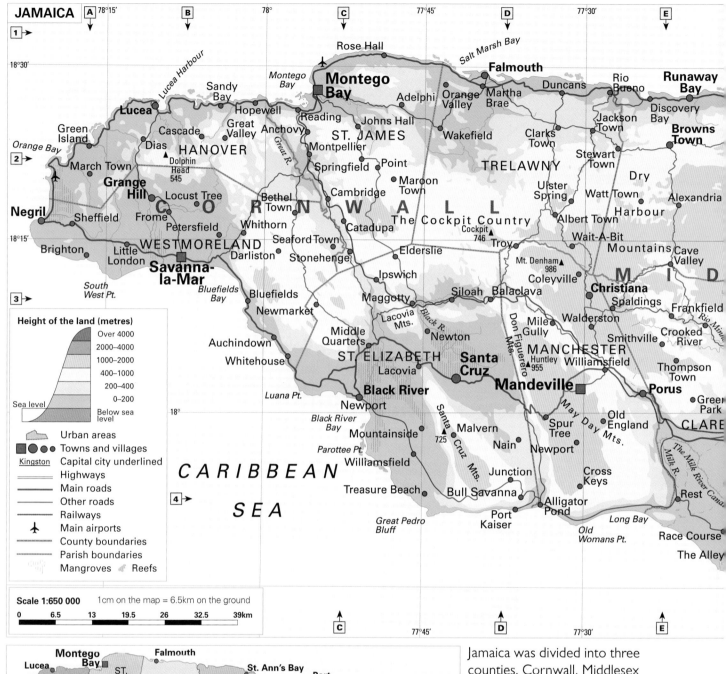

JAMAICA

A 78°15'   B   78°   C   77°45'   D   77°30'   E

18°30'

Rose Hall
Salt Marsh Bay
Falmouth
Montego Bay
Montego Bay
Rio Bueno   **Runaway Bay**
Sandy Bay
Hopewell   Reading
Adelphi   Orange Valley   Martha Brae   Duncans
Jackson Town   Discovery Bay   **Browns Town**
Lucea
Great Valley   Anchovy
Johns Hall
ST. JAMES
Wakefield   Clarks Town
Stewart Town   Dry
Green Island
Cascade
Montpellier
Springfield   Point
TRELAWNY
Ulster Spring   Watt Town
Alexandria
18°15'
Dias   **HANOVER**
Dolphin Head 545
Cambridge   Maroon Town
Albert Town   Harbour
March Town
Locust Tree
Bethel Town
Catadupa
The Cockpit Country
Wait-A-Bit   **M I D**
Grange Hill
C O R N W A L L
Cockpit 746   Troy
Mountains   Cave Valley
Negril
Sheffield   Frome
Whithorn
Seaford Town
Elderslie
Mt. Denham 986
Coleyville   **Christiana**
Brighton
Petersfield
WESTMORELAND
Darliston   Stonehenge
Ipswich
Siloah   Balaclava
Spaldings   Frankfield
Little London
Savanna-la-Mar
Bluefields Bay
Bluefields   Newmarket
Maggotty
Lacovia Mts.   Black R.
Mile Gully   Walderston
Smithville   Crooked River
South West Pt.
Auchindown
Middle Quarters   Newton
Don Figuerero Mts.
MANCHESTER
Huntley 955   Williamsfield
Thompson Town
Whitehouse
ST. ELIZABETH   **Santa Cruz**
Mandeville
Porus
Green Park
Luana Pt.
Lacovia
Black River
Newport
May Day Mts.
Old England
CLARE
18°
Black River Bay
Mountainside
Santa Cruz Mts. 725
Malvern   Nain
Spur Tree   Newport
Cross Keys
The Milk River Canal
Parottee Pt.
Williamsfield
Junction
Rest
Treasure Beach
Bull Savanna
Alligator Pond
Long Bay
Race Course
Great Pedro Bluff
Port Kaiser
Old Womans Pt.
The Alley

C A R I B B E A N
S E A

4

Height of the land (metres)
Over 4000
2000–4000
1000–2000
400–1000
200–400
0–200
Sea level
Below sea level

Urban areas
Towns and villages
Kingston   Capital city underlined
Highways
Main roads
Other roads
Railways
✈ Main airports
County boundaries
Parish boundaries
Mangroves   Reefs

Scale 1:650 000   1cm on the map = 6.5km on the ground
0   6.5   13   19.5   26   32.5   39km

C   77°45'   D   77°30'   E

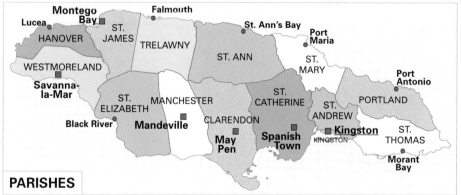

Montego Bay
Lucea
Falmouth
St. Ann's Bay
Port Maria
HANOVER   **ST. JAMES**   **TRELAWNY**   **ST. ANN**   **ST. MARY**
WESTMORELAND
Port Antonio
Savanna-la-Mar
ST. ELIZABETH   **MANCHESTER**   **ST. CATHERINE**   **PORTLAND**
Mandeville   **CLARENDON**   **ST. ANDREW**
Black River
May Pen   Spanish Town
Kingston   KINGSTON   **ST. THOMAS**
Morant Bay

PARISHES

Jamaica was divided into three counties, Cornwall, Middlesex and Surrey, in 1758 by the British governor. The counties now have little administrative function.
Kingston became the capital of Jamaica in 1872. Before this, the capital was Spanish Town. Jamaica is divided into 14 parishes, although Kingston and St. Andrew parishes are jointly administered.

CORNWALL   MIDDLESEX   SURREY

COUNTIES

NATIONAL SYMBOLS & COAT OF ARMS

Ackee, national fruit

Blue Mahoe, national tree

The Doctor Bird, national bird

Lignum Vitae, national flower

Coat of Arms

Renewal of downtown Kingston looking north from the Harbour

JAMAICA

Independence 1962
Capital: Kingston
Area: 10, 990 sq km
Population: 2,698,000 (2011 census)
Languages: English, Jamaican Creole
Sources of national income: sugar, bananas, tobacco, chemicals, coffee, tourism, alumina, bauxite, clothing

Legend

- Urban areas
- Parks/recreation areas/woodland
- Towns and villages
- Highways
- Main roads
- Other roads
- Railways

KINGSTON & PORTMORE

KINGSTON

Scale 1:100 000
1cm on the map = 1km on the ground

0   1   2   3   4km

COPYRIGHT PHILIP'S

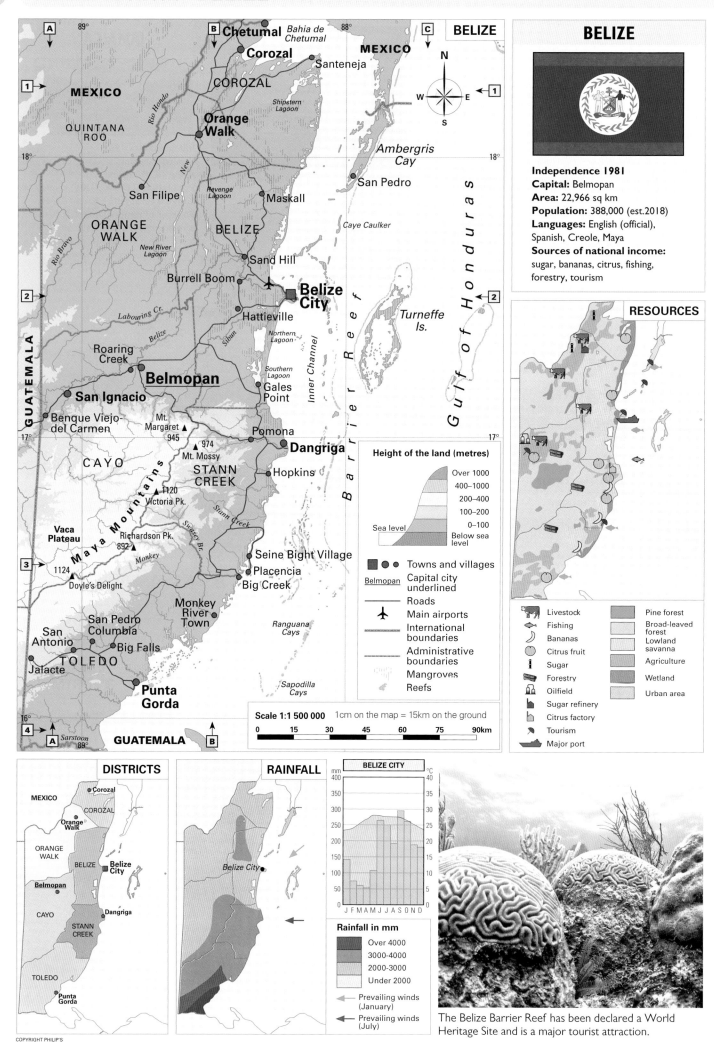

BELIZE

Independence 1981
Capital: Belmopan
Area: 22,966 sq km
Population: 388,000 (est.2018)
Languages: English (official), Spanish, Creole, Maya
Sources of national income: sugar, bananas, citrus, fishing, forestry, tourism

RESOURCES

Height of the land (metres)

| Over 1000 |
| 400–1000 |
| 200–400 |
| 100–200 |
| 0–100 |
| Below sea level |

Sea level

Towns and villages
Belmopan Capital city underlined
Roads
Main airports
International boundaries
Administrative boundaries
Mangroves
Reefs

Scale 1:1 500 000 1cm on the map = 15km on the ground
0 15 30 45 60 75 90km

Livestock
Fishing
Bananas
Citrus fruit
Sugar
Forestry
Oilfield
Sugar refinery
Citrus factory
Tourism
Major port

Pine forest
Broad-leaved forest
Lowland savanna
Agriculture
Wetland
Urban area

DISTRICTS

RAINFALL

BELIZE CITY

Rainfall in mm

| Over 4000 |
| 3000-4000 |
| 2000-3000 |
| Under 2000 |

Prevailing winds (January)
Prevailing winds (July)

The Belize Barrier Reef has been declared a World Heritage Site and is a major tourist attraction.

COPYRIGHT PHILIP'S

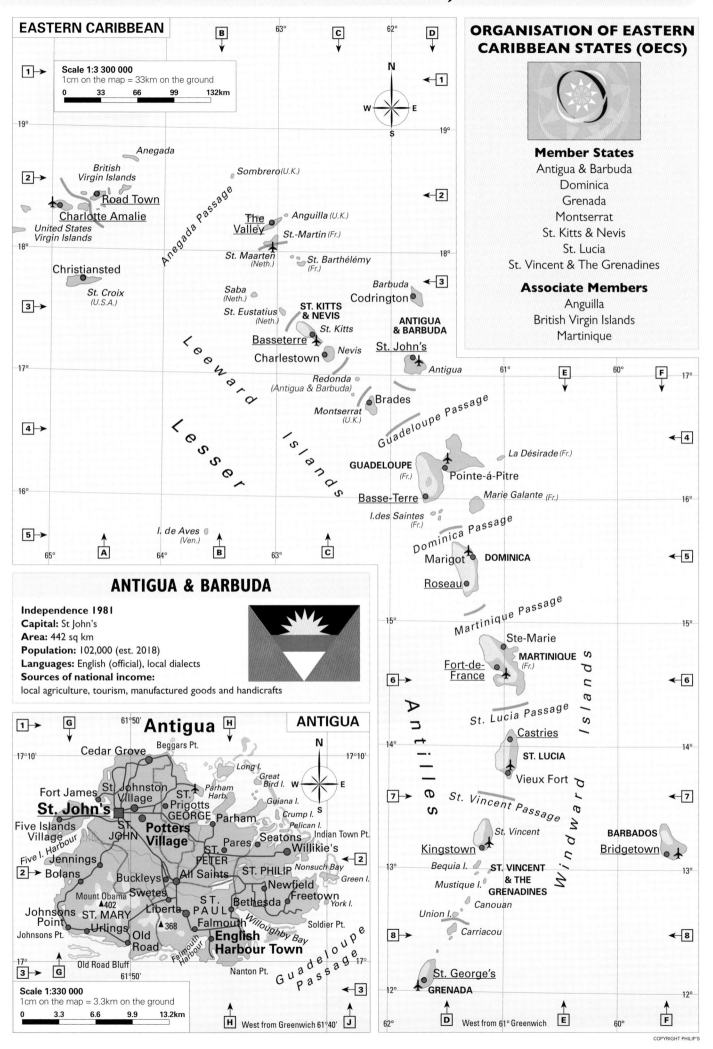

EASTERN CARIBBEAN

Scale 1:3 300 000
1cm on the map = 33km on the ground

0   33   66   99   132km

Anegada

British
Virgin Islands

Sombrero (U.K.)

Road Town

Charlotte Amalie

United States
Virgin Islands

Anguilla (U.K.)

The Valley

St.-Martin (Fr.)

Christiansted

St. Maarten (Neth.)

St. Barthélémy (Fr.)

St. Croix (U.S.A.)

Saba (Neth.)

Barbuda

Codrington

St. Eustatius (Neth.)

ST. KITTS & NEVIS

St. Kitts

ANTIGUA & BARBUDA

St. John's

Basseterre

Nevis

Charlestown

Antigua

Redonda
(Antigua & Barbuda)

Brades

Montserrat (U.K.)

La Désirade (Fr.)

Leeward Islands

GUADELOUPE (Fr.)

Pointe-á-Pitre

Basse-Terre

Marie Galante (Fr.)

Lesser

Antilles

I. des Saintes (Fr.)

I. de Aves (Ven.)

Dominica Passage

Marigot

DOMINICA

Roseau

Martinique Passage

Ste-Marie

MARTINIQUE (Fr.)

Fort-de-France

Windward Islands

St. Lucia Passage

Castries

ST. LUCIA

Vieux Fort

St. Vincent Passage

St. Vincent

Kingstown

BARBADOS

Bridgetown

Bequia I.

ST. VINCENT & THE GRENADINES

Mustique I.

Canouan

Union I.

Carriacou

St. George's

GRENADA

West from 61° Greenwich

ORGANISATION OF EASTERN CARIBBEAN STATES (OECS)

Member States
Antigua & Barbuda
Dominica
Grenada
Montserrat
St. Kitts & Nevis
St. Lucia
St. Vincent & The Grenadines

Associate Members
Anguilla
British Virgin Islands
Martinique

ANTIGUA & BARBUDA

Independence 1981
Capital: St John's
Area: 442 sq km
Population: 102,000 (est. 2018)
Languages: English (official), local dialects
Sources of national income:
local agriculture, tourism, manufactured goods and handicrafts

Antigua — ANTIGUA

Cedar Grove

Beggars Pt.

Long I.

Great Bird I.

Fort James

St Johnston Village

St. Prigotts

Parham Harb.

Guiana I.

St. John's

GEORGE

Parham

Crump I.

Five Islands Village

ST. JOHN

Potters Village

Pares

Pelican I.

Indian Town Pt.

Seatons

Five I. Harbour

Jennings

ST. PETER

Willikie's

Bolans

Buckleys

All Saints

ST. PHILIP

Nonsuch Bay

Swetes

Mount Obama ▲402

ST. MARY

Liberta

ST. PAUL

Bethesda

Newfield

Green I.

Freetown

York I.

Johnsons Point

Urlings

▲368

Falmouth

Willoughby Bay

Soldier Pt.

Johnsons Pt.

Old Road

English Harbour Town

Falmouth Harbour

Old Road Bluff

Nanton Pt.

Guadeloupe Passage

Scale 1:330 000
1cm on the map = 3.3km on the ground

0   3.3   6.6   9.9   13.2km

West from Greenwich 61°40'

COPYRIGHT PHILIP'S

ST. KITTS & NEVIS

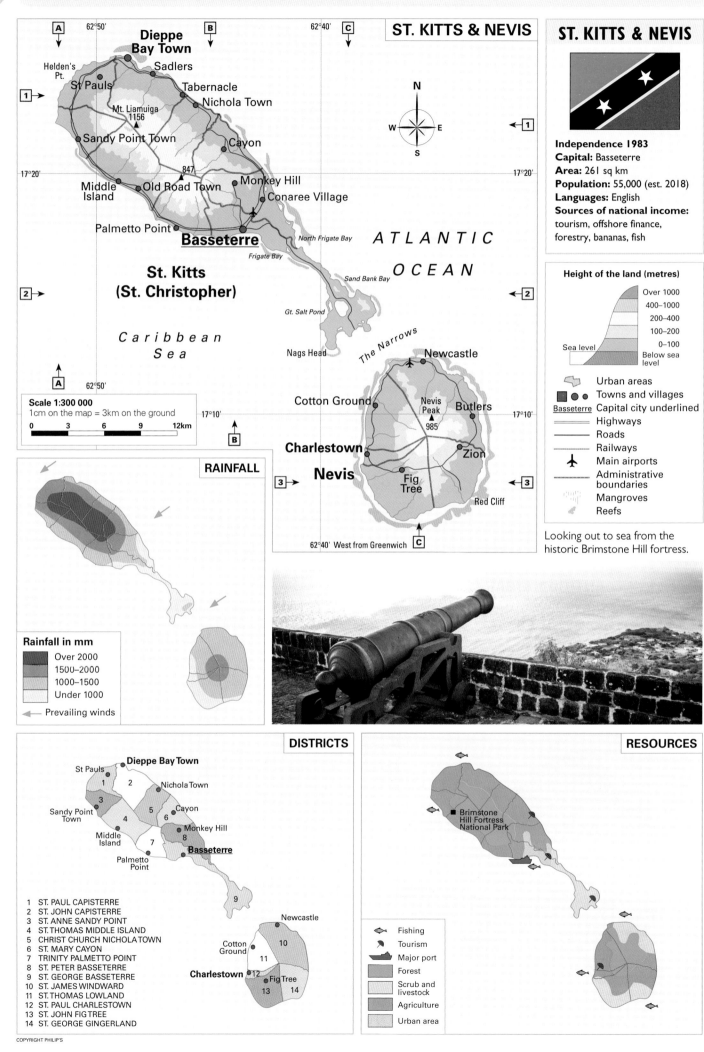

A | 62°50' | B | 62°40' | C

Dieppe Bay Town
Helden's Pt.
Sadlers
St Pauls
Tabernacle
Nichola Town
Mt. Liamuiga 1156 ▲
Sandy Point Town
Cayon
17°20'
847 ▲
Middle Island
Old Road Town
Monkey Hill
Palmetto Point
Conaree Village
Basseterre
North Frigate Bay
Frigate Bay

St. Kitts (St. Christopher)

Caribbean Sea

Scale 1:300 000
1cm on the map = 3km on the ground
0 3 6 9 12km

ATLANTIC OCEAN

Sand Bank Bay
Gt. Salt Pond
Nags Head
The Narrows
Newcastle
Cotton Ground
Nevis Peak ▲ 985
Butlers
Charlestown
Nevis
Fig Tree
Zion
Red Cliff
62°40' West from Greenwich C

ST. KITTS & NEVIS

Independence 1983
Capital: Basseterre
Area: 261 sq km
Population: 55,000 (est. 2018)
Languages: English
Sources of national income: tourism, offshore finance, forestry, bananas, fish

Height of the land (metres)

Over 1000
400–1000
200–400
100–200
0–100
Sea level
Below sea level

Urban areas
Towns and villages
Basseterre Capital city underlined
Highways
Roads
Railways
✈ Main airports
Administrative boundaries
Mangroves
Reefs

Looking out to sea from the historic Brimstone Hill fortress.

RAINFALL

Rainfall in mm

Over 2000
1500–2000
1000–1500
Under 1000
→ Prevailing winds

DISTRICTS

Dieppe Bay Town
St Pauls
1
2
Nichola Town
3
Sandy Point Town
4
5
6
Cayon
Middle Island
7
8
Monkey Hill
Palmetto Point
Basseterre
9

Newcastle
Cotton Ground
11
10
12
Fig Tree
Charlestown
13
14

1 ST. PAUL CAPISTERRE
2 ST. JOHN CAPISTERRE
3 ST. ANNE SANDY POINT
4 ST. THOMAS MIDDLE ISLAND
5 CHRIST CHURCH NICHOLA TOWN
6 ST. MARY CAYON
7 TRINITY PALMETTO POINT
8 ST. PETER BASSETERRE
9 ST. GEORGE BASSETERRE
10 ST. JAMES WINDWARD
11 ST. THOMAS LOWLAND
12 ST. PAUL CHARLESTOWN
13 ST. JOHN FIG TREE
14 ST. GEORGE GINGERLAND

RESOURCES

Brimstone Hill Fortress National Park

Fishing
Tourism
Major port
Forest
Scrub and livestock
Agriculture
Urban area

DOMINICA

Independence 1978
Capital: Roseau
Area: 751 sq km
Population: 74,000 (est. 2018)
Languages: English (official) French patois
Sources of national income: bananas, citrus, coconut oil, soap, ecotourism

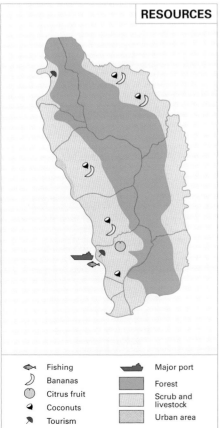

Dominica farmers produce food to export to neighbouring islands such as dasheen, yam, tania, plantain and citrus.

RESOURCES

Fishing
Bananas
Citrus fruit
Coconuts
Tourism

Major port
Forest
Scrub and livestock
Urban area

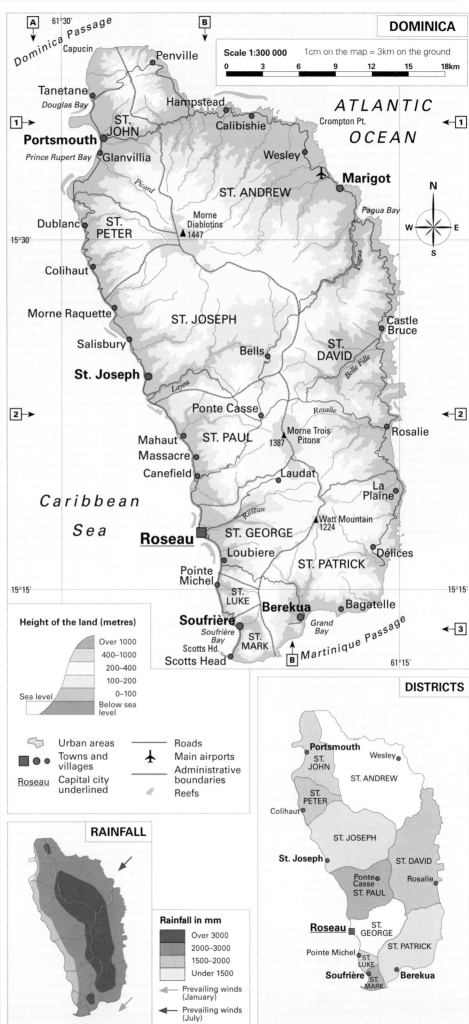

DOMINICA

Scale 1:300 000   1cm on the map = 3km on the ground

0   3   6   9   12   15   18km

61°30'

Dominica Passage

Capucin
Penville
Tanetane
Douglas Bay
Hampstead
ST. JOHN
Calibishie
Crompton Pt.
Portsmouth
Prince Rupert Bay
Glanvillia
Wesley
Picard
Marigot
ST. ANDREW
Dublanc
ST. PETER
Morne Diablotins ▲ 1447
Pagua Bay
15°30'
Colihaut
Pagua
Morne Raquette
ST. JOSEPH
Castle Bruce
Salisbury
Bells
ST. DAVID
St. Joseph
Belle Fille
Layou
Ponte Casse
Rosalie
Mahaut
ST. PAUL
Morne Trois Pitons 1387
Rosalie
Massacre
Canefield
Laudat
La Plaine
C a r i b b e a n   S e a
Roseau
Watt Mountain 1224
Roseau
ST. GEORGE
Loubiere
Délices
Pointe Michel
ST. PATRICK
15°15'
ST. LUKE
Berekua
Bagatelle
Soufrière
Soufrière Bay
Grand Bay
Scotts Hd.
ST. MARK
Martinique Passage
Scotts Head
61°15'

ATLANTIC OCEAN

N
W   E
S

Height of the land (metres)

Over 1000
400–1000
200–400
100–200
0–100
Sea level
Below sea level

Urban areas
Towns and villages
Roseau Capital city underlined

Roads
✈ Main airports
Administrative boundaries
Reefs

RAINFALL

Rainfall in mm

Over 3000
2000–3000
1500–2000
Under 1500

Prevailing winds (January)
Prevailing winds (July)

DISTRICTS

Portsmouth
ST. JOHN
Wesley
ST. ANDREW
ST. PETER
Colihaut
ST. JOSEPH
St. Joseph
ST. DAVID
Ponte Casse
Rosalie
ST. PAUL
Roseau
ST. GEORGE
Pointe Michel
ST. PATRICK
ST. LUKE
Soufrière
ST. MARK
Berekua

COPYRIGHT PHILIP'S

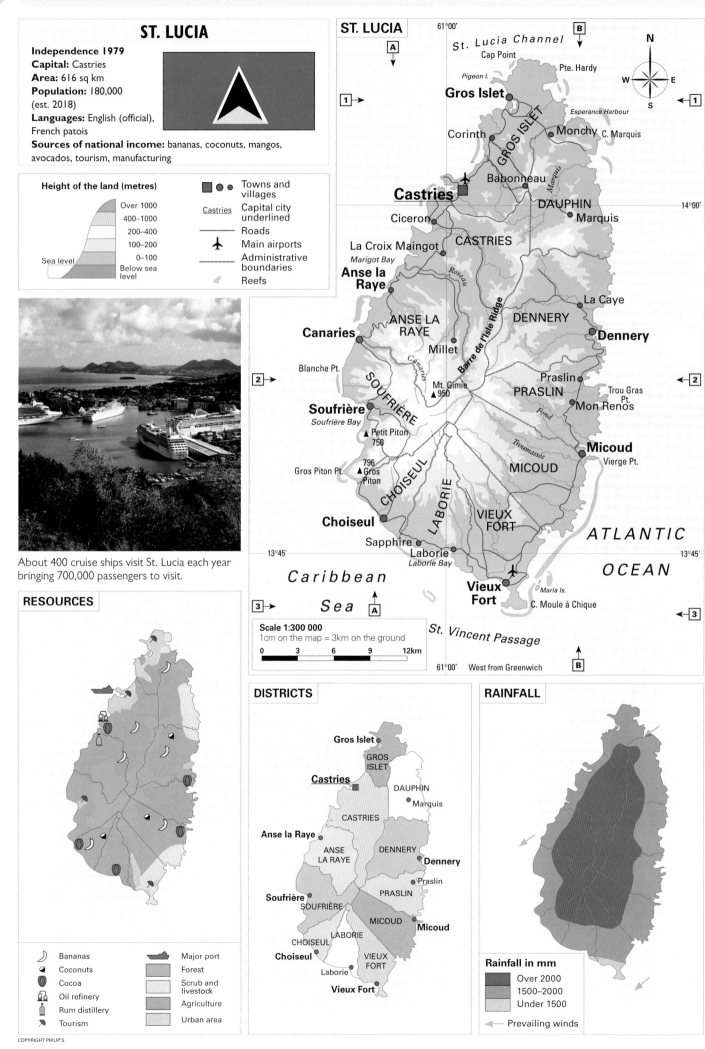

ST. LUCIA

Independence 1979
Capital: Castries
Area: 616 sq km
Population: 180,000
(est. 2018)
Languages: English (official),
French patois
Sources of national income: bananas, coconuts, mangos,
avocados, tourism, manufacturing

Height of the land (metres)

Over 1000
400–1000
200–400
100–200
0–100
Sea level
Below sea
level

Towns and villages
Castries Capital city underlined
Roads
✈ Main airports
Administrative boundaries
Reefs

About 400 cruise ships visit St. Lucia each year
bringing 700,000 passengers to visit.

RESOURCES

Bananas
Coconuts
Cocoa
Oil refinery
Rum distillery
Tourism
Major port
Forest
Scrub and livestock
Agriculture
Urban area

COPYRIGHT PHILIP'S

ST. LUCIA

61°00'

St. Lucia Channel
Cap Point
Pigeon I.
Pte. Hardy
Gros Islet
Esperance Harbour
Corinth
Monchy C. Marquis
Babonneau
Castries DAUPHIN
Ciceron Marquis
La Croix Maingot CASTRIES
Marigot Bay
Anse la Raye
ANSE LA RAYE
Canaries
Millet
Blanche Pt.
Mt. Gimie ▲ 950
Soufrière
Soufrière Bay
▲ Petit Piton 750
Gros Piton Pt. 796 ▲ Gros Piton
Choiseul
Sapphire
Laborie
Laborie Bay
Vieux Fort
C. Moule à Chique
Maria Is.
La Caye
DENNERY
Dennery
Praslin
PRASLIN Trou Gras Pt.
Mon Renos
Micoud
Vierge Pt.
MICOUD
VIEUX FORT
LABORIE

GROS ISLET
Barre de l'Isle Ridge
SOUFRIÈRE
CHOISEUL

Caribbean Sea

ATLANTIC OCEAN

14°00'
13°45'

Scale 1:300 000
1cm on the map = 3km on the ground
0 3 6 9 12km

St. Vincent Passage

61°00' West from Greenwich

DISTRICTS

Gros Islet
GROS ISLET
Castries
DAUPHIN
Marquis
CASTRIES
Anse la Raye
ANSE LA RAYE
DENNERY
Dennery
Praslin
PRASLIN
Soufrière
SOUFRIÈRE
MICOUD
Micoud
LABORIE
CHOISEUL
Choiseul
Laborie
VIEUX FORT
Vieux Fort

RAINFALL

Rainfall in mm
Over 2000
1500–2000
Under 1500
Prevailing winds

ST. VINCENT & THE GRENADINES

ST. VINCENT & THE GRENADINES

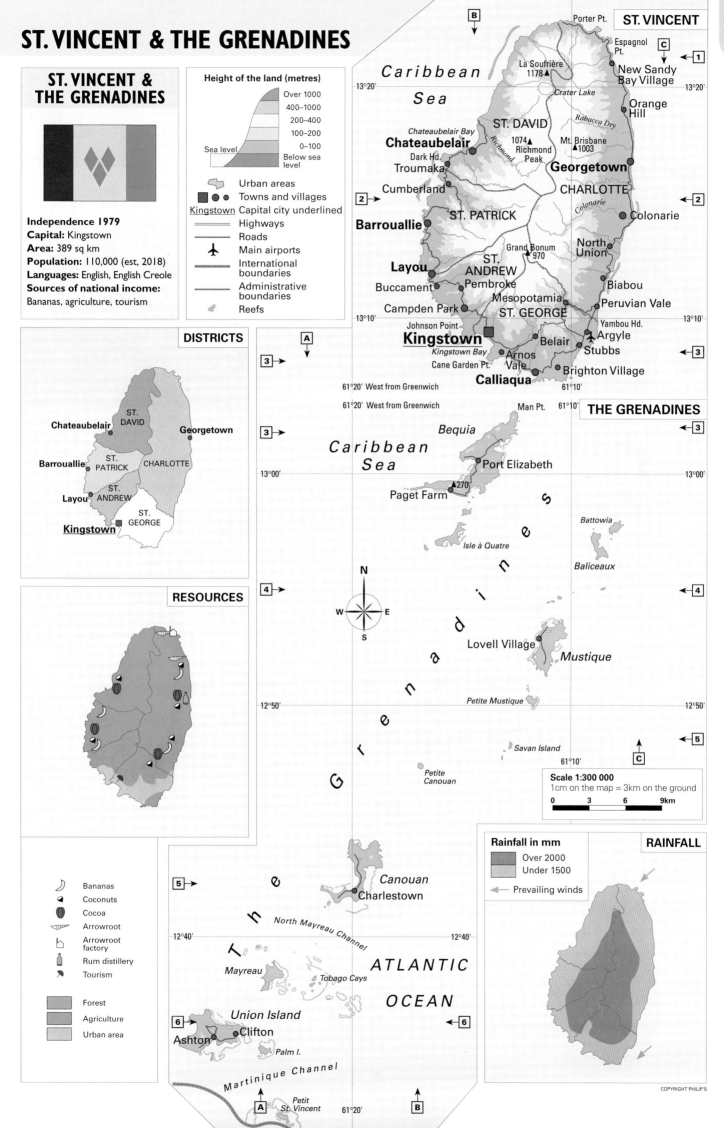

Independence 1979
Capital: Kingstown
Area: 389 sq km
Population: 110,000 (est, 2018)
Languages: English, English Creole
Sources of national income:
Bananas, agriculture, tourism

Height of the land (metres)

Over 1000
400–1000
200–400
100–200
0–100
Sea level
Below sea level

Urban areas
Towns and villages
Kingstown Capital city underlined
Highways
Roads
Main airports
International boundaries
Administrative boundaries
Reefs

DISTRICTS

Chateaubelair
ST. DAVID
Georgetown
Barrouallie
ST. PATRICK
CHARLOTTE
Layou
ST. ANDREW
Kingstown
ST. GEORGE

RESOURCES

Bananas
Coconuts
Cocoa
Arrowroot
Arrowroot factory
Rum distillery
Tourism

Forest
Agriculture
Urban area

ST. VINCENT

Porter Pt.
Espagnol Pt.
New Sandy Bay Village
Orange Hill

Caribbean Sea

La Soufrière 1178▲
Crater Lake
Rabacca Dry
ST. DAVID
Chateaubelair Bay
Chateaubelair
Dark Hd.
Troumaka
Richmond
1074▲ Richmond Peak
Mt. Brisbane ▲1003
Georgetown
Cumberland
CHARLOTTE
Colonarie
ST. PATRICK
Colonarie
Barrouallie
North Union
Grand Bonum ▲970
Layou
ST. ANDREW
Buccament
Pembroke
Biabou
Campden Park
Mesopotamia
Peruvian Vale
ST. GEORGE
Johnson Point
Yambou Hd.
Kingstown
Belair
Argyle
Kingstown Bay
Arnos Vale
Stubbs
Cane Garden Pt.
Calliaqua
Brighton Village

13°20'
13°10'
61°20' West from Greenwich

THE GRENADINES

Man Pt.

Bequia
Port Elizabeth
Caribbean Sea
▲270
Paget Farm

Battowia
Isle à Quatre
Baliceaux

Lovell Village
Mustique

Petite Mustique

Savan Island
Petite Canouan

13°00'
12°50'

Scale 1:300 000
1cm on the map = 3km on the ground
0   3   6   9km

Canouan
Charlestown

North Mayreau Channel

The

ATLANTIC OCEAN

Mayreau
Tobago Cays

Union Island
Clifton
Ashton
Palm I.

Martinique Channel

Petit St. Vincent

12°40'
61°20'

RAINFALL

Rainfall in mm
Over 2000
Under 1500
Prevailing winds

COPYRIGHT PHILIP'S

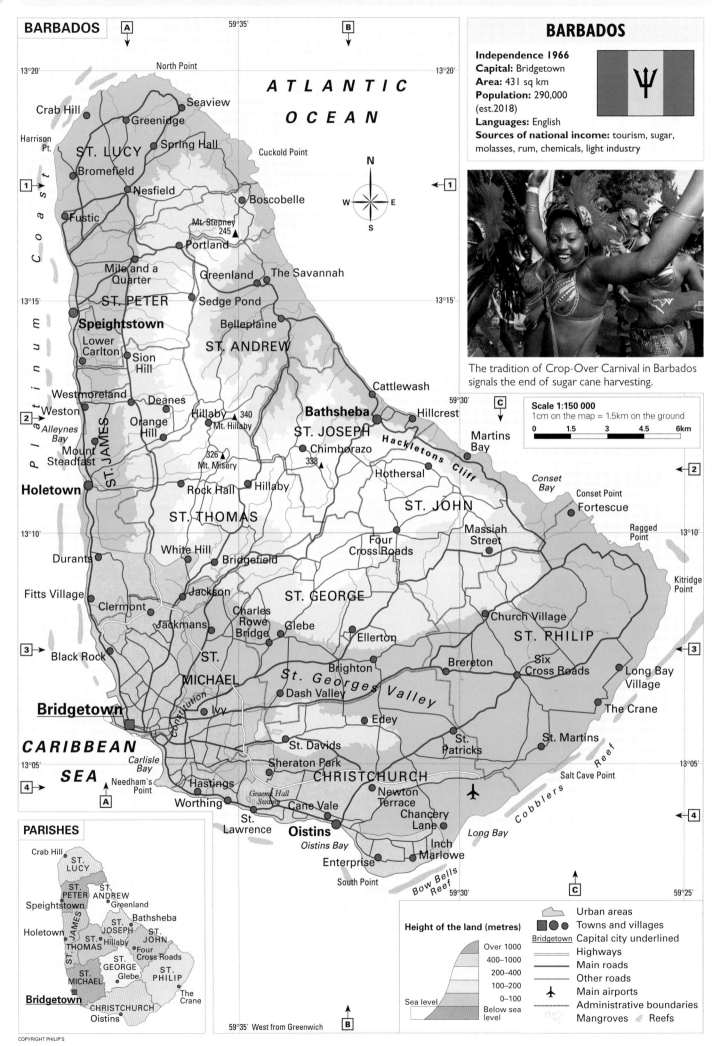

BARBADOS

BARBADOS

Independence 1966
Capital: Bridgetown
Area: 431 sq km
Population: 290,000
(est.2018)
Languages: English
Sources of national income: tourism, sugar, molasses, rum, chemicals, light industry

The tradition of Crop-Over Carnival in Barbados signals the end of sugar cane harvesting.

Scale 1:150 000
1cm on the map = 1.5km on the ground

0   1.5   3   4.5   6km

Map labels

ATLANTIC OCEAN

North Point
Crab Hill
Seaview
Greenidge
Harrison Pt.
Spring Hall
ST. LUCY
Bromefield
Nesfield
Cuckold Point
Fustic
Boscobelle
Mt. Stepney 245
Portland
Mile and a Quarter
Greenland
The Savannah
ST. PETER
Sedge Pond
Speightstown
Belleplaine
Lower Carlton
Sion Hill
ST. ANDREW
Westmoreland
Deanes
Cattlewash
Weston
Hillaby 340
Bathsheba
Alleynes Bay
Orange Hill
Mt. Hillaby
ST. JOSEPH
Hillcrest
Mount Steadfast
326 Mt. Misery
338
Chimborazo
Martins Bay
Holetown
Rock Hall
Hillaby
Hothersal
Conset Bay
Conset Point
Fortescue
ST. THOMAS
ST. JOHN
Massiah Street
Ragged Point
Durants
White Hill
Bridgefield
Four Cross Roads
Kitridge Point
Fitts Village
Jackson
ST. GEORGE
Church Village
Clermont
Charles Rowe Bridge
Glebe
Ellerton
ST. PHILIP
Jackmans
Brereton
Six Cross Roads
Long Bay Village
Black Rock
ST. MICHAEL
Brighton
St. Georges Valley
Dash Valley
The Crane
Bridgetown
Ivy
Edey
St. Patricks
St. Martins
CARIBBEAN SEA
Carlisle Bay
Needham's Point
St. Davids
Sheraton Park
CHRISTCHURCH
Salt Cave Point
Hastings
Graeme Hall Swamp
Newton Terrace
Long Bay
Worthing
Cane Vale
Chancery Lane
Cobblers Reef
St. Lawrence
Oistins
Inch Marlowe
Oistins Bay
Enterprise
Bow Bells Reef
South Point

Platinum Coast
Hackletons Cliff
Constitution

PARISHES

Crab Hill
ST. LUCY
ST. PETER
ST. ANDREW
Speightstown
Greenland
Bathsheba
Holetown
ST. JAMES
ST. JOSEPH
ST. JOHN
ST. THOMAS
Hillaby
Four Cross Roads
ST. GEORGE
ST. PHILIP
ST. MICHAEL
Glebe
Bridgetown
The Crane
CHRISTCHURCH
Oistins

Legend

Height of the land (metres)

Over 1000
400–1000
200–400
100–200
0–100
Sea level
Below sea level

Urban areas
Towns and villages
Bridgetown Capital city underlined
Highways
Main roads
Other roads
Main airports
Administrative boundaries
Mangroves   Reefs

59°35'
59°30'
59°25'
13°20'
13°15'
13°10'
13°05'
59°35' West from Greenwich

GRENADA

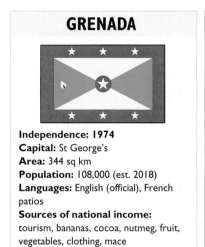

Independence: 1974
Capital: St George's
Area: 344 sq km
Population: 108,000 (est. 2018)
Languages: English (official), French patios
Sources of national income:
tourism, bananas, cocoa, nutmeg, fruit, vegetables, clothing, mace

Height of the land (metres)

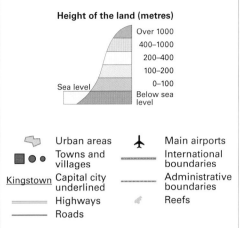

Over 1000
400–1000
200–400
100–200
0–100
Below sea level

Sea level

Urban areas
Towns and villages
Kingstown Capital city underlined
Highways
Roads

✈ Main airports
International boundaries
Administrative boundaries
Reefs

CARRIACOU

Same scale as Grenada

61°30'

Petit St. Vincent
Gun Pt.
291 ▲ **Windward**
Petite Martinique

12°30'

Hillsborough Bay

Caribbean Sea

✈ **Hillsborough**

Tyrrel Bay ● **Hermitage**

Southwest Pt.

Carriacou (Grenada)

Saline I.

ATLANTIC OCEAN

Frigate I.

Large I.

61°30'

GRENADA

A 61°40' B Diamond I.
Les Tantes
Ronde Island
Caille I.

Caribbean Sea

N W E S

Tanga Langua
Green I.
Sauteurs
Bedford Pt.
Grenada Bay
Victoria
ST. PATRICK
River Sallee
Tricolar
ST. MARK
Artiste Pt.
840 ▲ Tivoli
Gouyave
Mt. St. Catherine
Pearls
Grand **ST. JOHN**
Roy
Great River Bay
Concord
Great River
Grenville
ST. ANDREW
Telescope Pt.
Grenville Bay
Birch Grove

12°10'

Moliniere Pt.
702 ▲ St. Francis
Mt. Sinai
ATLANTIC

St. George's
ST. GEORGE
Great Bacolet Bay

Grand Anse Bay
St. Pauls
ST. DAVID
St. Davids
OCEAN

Morne Rouge
Corinth
Requin Bay

Pt. Salines
12°00'
Lance aux Epines
Hog I.
Pt. of Fort Jeudy
Glover I. Prickly Pt.
Calivigny I.
West from Greenwich 61°40'

Scale 1:300 000
1cm on the map = 3km on the ground
0 3 6 9 12km

RESOURCES

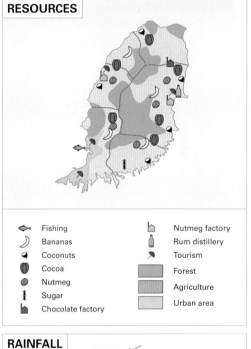

🐟 Fishing
🍌 Bananas
Coconuts
Cocoa
Nutmeg
Sugar
Chocolate factory

Nutmeg factory
Rum distillery
Tourism
Forest
Agriculture
Urban area

RAINFALL

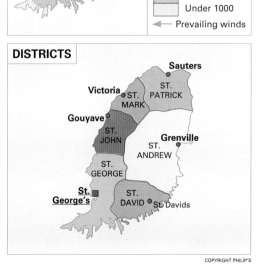

Rainfall in mm
Over 3000
2000–3000
1500–2000
1000–1500
Under 1000
⬅ Prevailing winds

DISTRICTS

Sauters
Victoria
ST. MARK
ST. PATRICK
Gouyave
ST. JOHN
Grenville
ST. ANDREW
ST. GEORGE
St. George's
ST. DAVID St Davids

The picturesque Carenage of Grenada's capital St Georges. Increasingly Grenada has become a centre for yachting and the harbour hosts a marina.

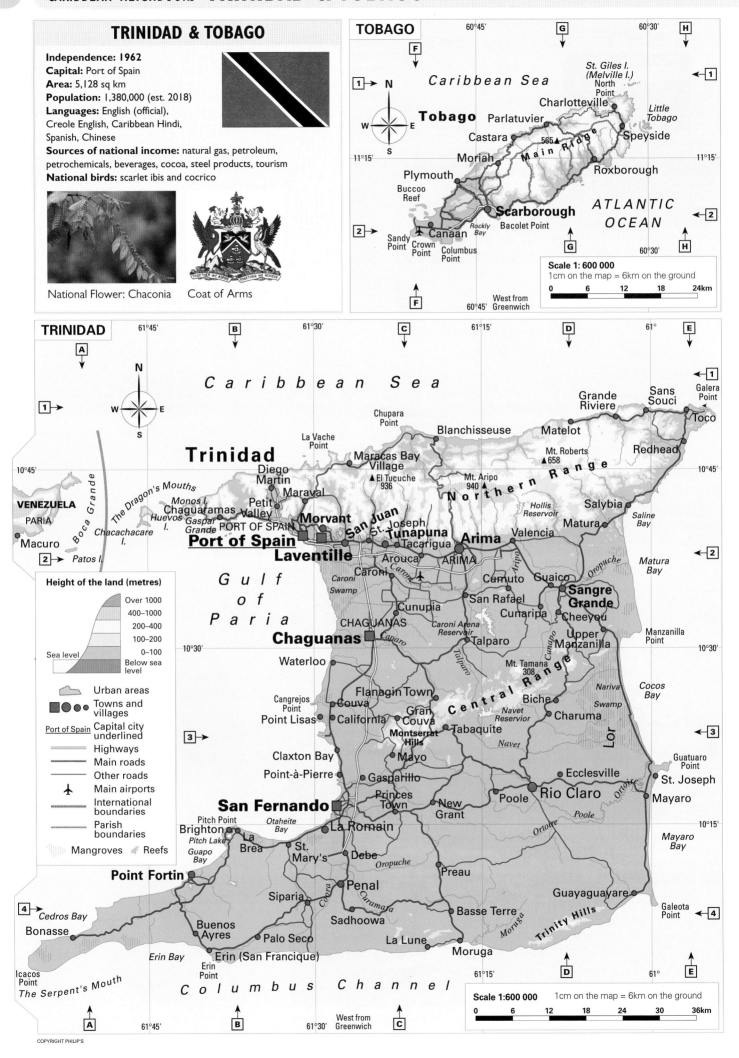

TRINIDAD & TOBAGO

Independence: 1962
Capital: Port of Spain
Area: 5,128 sq km
Population: 1,380,000 (est. 2018)
Languages: English (official),
Creole English, Caribbean Hindi,
Spanish, Chinese
Sources of national income: natural gas, petroleum,
petrochemicals, beverages, cocoa, steel products, tourism
National birds: scarlet ibis and cocrico

National Flower: Chaconia   Coat of Arms

TOBAGO

Scale 1: 600 000
1cm on the map = 6km on the ground
0   6   12   18   24km

West from
Greenwich

TRINIDAD

Height of the land (metres)

Over 1000
400–1000
200–400
100–200
0–100
Sea level
Below sea level

Urban areas
Towns and villages
Port of Spain Capital city underlined
Highways
Main roads
Other roads
Main airports
International boundaries
Parish boundaries
Mangroves   Reefs

Scale 1:600 000   1cm on the map = 6km on the ground
0   6   12   18   24   30   36km

West from
Greenwich

COPYRIGHT PHILIP'S

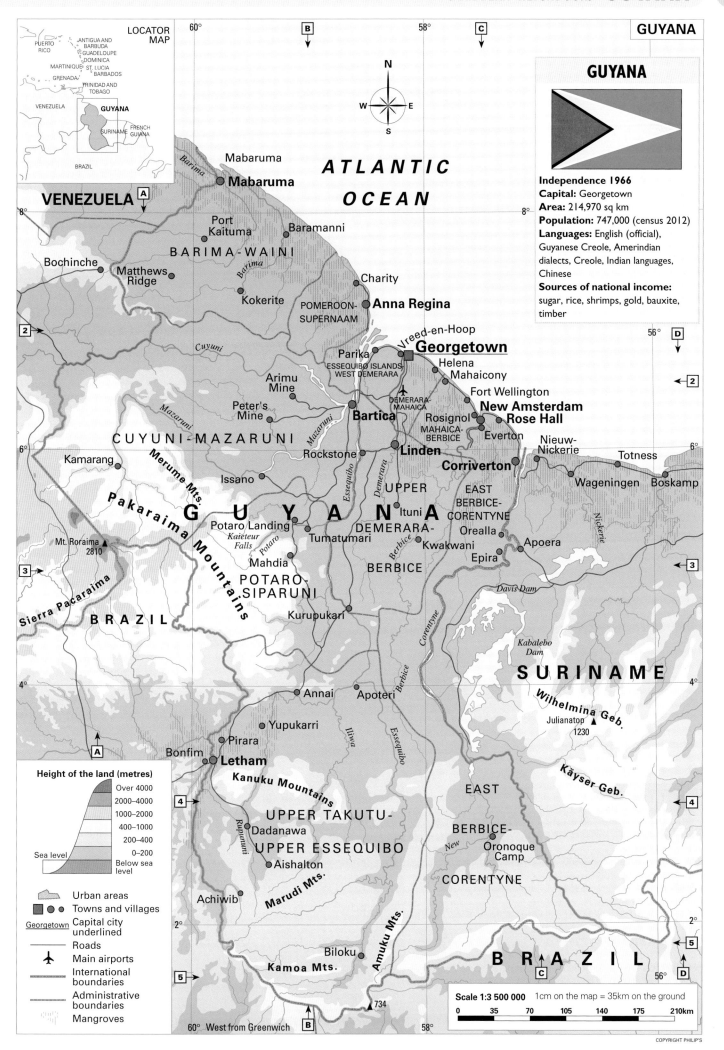

LOCATOR MAP

PUERTO RICO
ANTIGUA AND BARBUDA
GUADELOUPE
DOMINICA
MARTINIQUE ST. LUCIA
BARBADOS
GRENADA
TRINIDAD AND TOBAGO
VENEZUELA
GUYANA
SURINAME FRENCH GUIANA
BRAZIL

GUYANA

Independence 1966
Capital: Georgetown
Area: 214,970 sq km
Population: 747,000 (census 2012)
Languages: English (official), Guyanese Creole, Amerindian dialects, Creole, Indian languages, Chinese
Sources of national income: sugar, rice, shrimps, gold, bauxite, timber

ATLANTIC OCEAN

VENEZUELA

Mabaruma
Mabaruma

Port Kaituma
Baramanni

BARIMA-WAINI

Bochinche
Matthews Ridge
Charity
Kokerite
POMEROON-SUPERNAAM
Anna Regina

Vreed-en-Hoop
Georgetown
Parika
Helena
ESSEQUIBO ISLANDS WEST DEMERARA
Mahaicony
Arimu Mine
Fort Wellington
New Amsterdam
Peter's Mine
DEMERARA-MAHAICA
Rosignol
Rose Hall
Bartica
MAHAICA-BERBICE
Everton
Nieuw-Nickerie
CUYUNI-MAZARUNI
Rockstone
Linden
Corriverton
Totness

Kamarang
G U Y A N A
UPPER
EAST BERBICE-CORENTYNE
Wageningen
Boskamp
Issano
Ituni
Potaro Landing
Kaieteur Falls
Tumatumari
DEMERARA-
Orealla
Mt. Roraima 2810
Mahdia
BERBICE
Kwakwani
Apoera
Epira
POTARO-SIPARUNI
Kurupukari
Davis Dam

Sierra Pacaraima
B R A Z I L
Kabalebo Dam

S U R I N A M E

Annai
Apoteri
Wilhelmina Geb.
Julianatop 1230

Yupukarri
Pirara
Bonfim
Letham
Kanuku Mountains
EAST
Kayser Geb.

UPPER TAKUTU-
Dadanawa
BERBICE-
Oronoque Camp
UPPER ESSEQUIBO
Aishalton
CORENTYNE

Achiwib
Marudi Mts.

Biloku
B R A Z I L
Kamoa Mts.
734

Barima
Barima
Cuyuni
Mazaruni
Mazaruni
Essequibo
Demerara
Berbice
Nickerie
Corentyne
Berbice
Ireng
Essequibo
Rupununi
New
Amuku Mts.

Pakaraima Mountains
Merume Mts.

Height of the land (metres)

Over 4000
2000–4000
1000–2000
400–1000
200–400
0–200
Sea level
Below sea level

Urban areas
Towns and villages
Georgetown Capital city underlined
Roads
Main airports
International boundaries
Administrative boundaries
Mangroves

Scale 1:3 500 000 1cm on the map = 35km on the ground

0   35   70   105   140   175   210km

60° West from Greenwich

COPYRIGHT PHILIP'S

FACT FILE

Continent	Area '000 sq km	Coldest place °C		Hottest place °C		Wettest place average annual rainfall, mm		Driest place average annual rainfall, mm
ASIA	44,500	Oymyakon, Russia -70°C	A1	Tirat Zevi, Israel 54°C	B1	Mawsynram, India 11,870	C1	Aden, Yemen 46
AFRICA	30,302	Ifrane, Morocco -24°C	A2	Kebili, Tunisia 55°C	B2	Debundscha, Cameroon 10,290	C2	Wadi Haifa, Sudan 2
NORTH AMERICA	24,241	Snag, Canada -63°C	A3	Death Valley, California 57°C	B3	Henderson Lake, Canada 6,500	C3	Bataques, Mexico 30
SOUTH AMERICA	17,793	Sarmiento, Argentina -33°C	A4	Rivadavia, Argentina 49°C	B4	Quibdó, Colombia 8,990	C4	Quillagua, Chile 0.6
ANTARCTICA	14,000	Vostok -89°C	A5	Vanda Station 15°C	B5			
EUROPE	9,957	Ust'Shchugor, Russia -55°C	A6	Seville, Spain 50°C	B6	Crkvice, Montenegro 4,650	C5	Astrakhan, Russia 160
AUSTRALIA	8,557	Charlotte Pass, Australia -22°C	A7	Cloncurry, Australia 53°C	B7	Tully, Australia 4,550	C6	Mulka, Australia 100

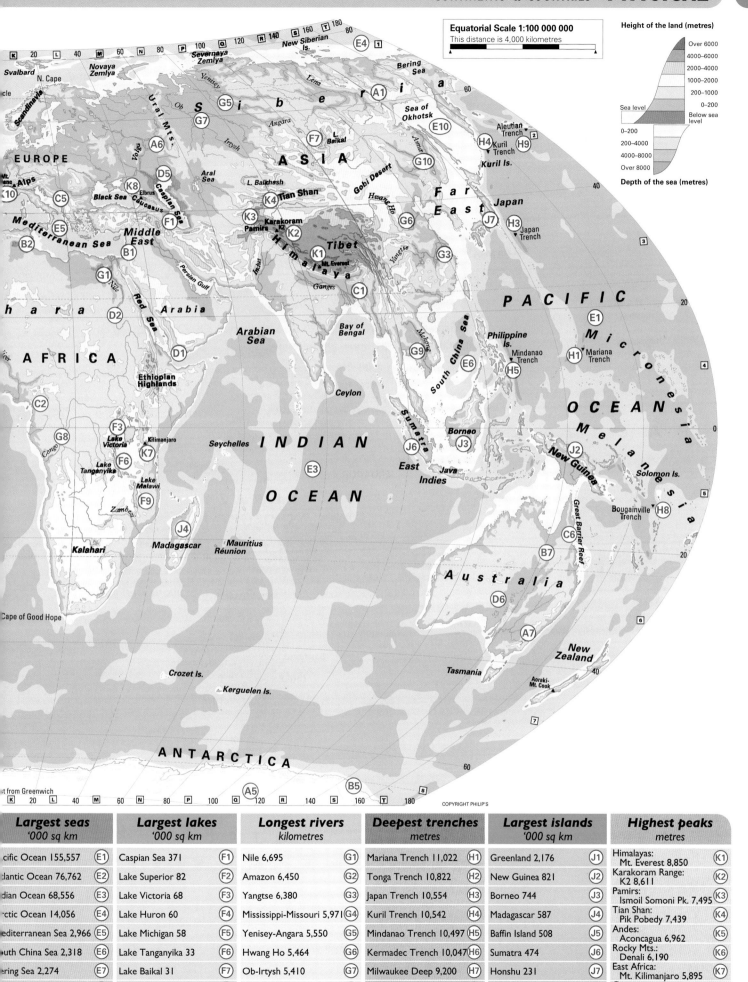

Equatorial Scale 1:100 000 000
This distance is 4,000 kilometres

Height of the land (metres)

	Over 6000
	4000–6000
	2000–4000
	1000–2000
	200–1000
	0–200
Sea level	Below sea level

0–200	
200–4000	
4000–8000	
Over 8000	

Depth of the sea (metres)

COPYRIGHT PHILIP'S

Largest seas '000 sq km		Largest lakes '000 sq km		Longest rivers kilometres		Deepest trenches metres		Largest islands '000 sq km		Highest peaks metres	
cific Ocean 155,557	E1	Caspian Sea 371	F1	Nile 6,695	G1	Mariana Trench 11,022	H1	Greenland 2,176	J1	Himalayas: Mt. Everest 8,850	K1
lantic Ocean 76,762	E2	Lake Superior 82	F2	Amazon 6,450	G2	Tonga Trench 10,822	H2	New Guinea 821	J2	Karakoram Range: K2 8,611	K2
dian Ocean 68,556	E3	Lake Victoria 68	F3	Yangtse 6,380	G3	Japan Trench 10,554	H3	Borneo 744	J3	Pamirs: Ismoil Somoni Pk. 7,495	K3
ctic Ocean 14,056	E4	Lake Huron 60	F4	Mississippi-Missouri 5,971	G4	Kuril Trench 10,542	H4	Madagascar 587	J4	Tian Shan: Pik Pobedy 7,439	K4
editerranean Sea 2,966	E5	Lake Michigan 58	F5	Yenisey-Angara 5,550	G5	Mindanao Trench 10,497	H5	Baffin Island 508	J5	Andes: Aconcagua 6,962	K5
uth China Sea 2,318	E6	Lake Tanganyika 33	F6	Hwang Ho 5,464	G6	Kermadec Trench 10,047	H6	Sumatra 474	J6	Rocky Mts.: Denali 6,190	K6
ering Sea 2,274	E7	Lake Baikal 31	F7	Ob-Irtysh 5,410	G7	Milwaukee Deep 9,200	H7	Honshu 231	J7	East Africa: Mt. Kilimanjaro 5,895	K7
ribbean Sea 1,942	E8	Great Bear Lake 31	F8	Congo 4,670	G8	Bouganville Trench 9,140	H8	Great Britain 230	J8	Caucasus: Elbrus 5,640	K8
ulf of Mexico 1,813	E9	Lake Malawi 30	F9	Mekong 4,500	G9	Aleutian Trench 7,822	H9	Victoria Island 212	J9	Antarctica: Vinson Massif 4,897	K9
a of Okhotsk 1,528	E10	Great Slave Lake 29	F10	Amur 4,442	G10	South Sandwich Island Trench 7,235	H10	Ellesmere Island 197	J10	Alps: Mt. Blanc 4,808	K10

COPYRIGHT PHILIP'S

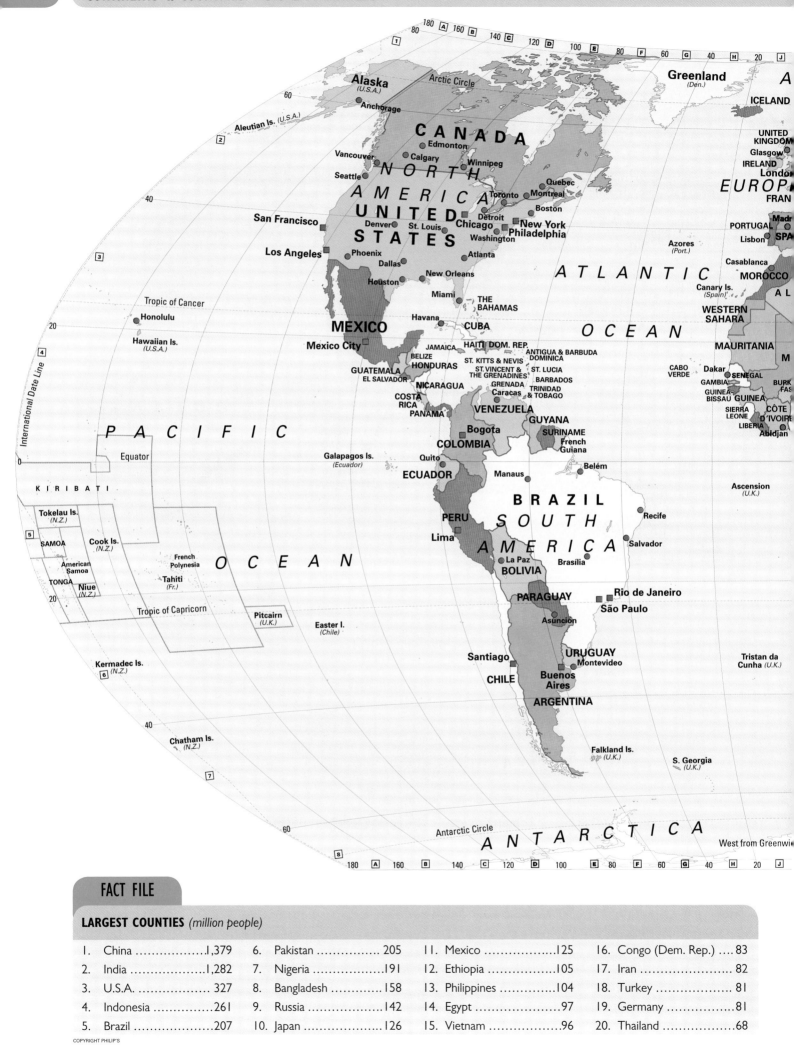

FACT FILE

LARGEST COUNTIES *(million people)*

1.	China 1,379	6.	Pakistan 205	11.	Mexico 125	16.	Congo (Dem. Rep.) 83
2.	India 1,282	7.	Nigeria 191	12.	Ethiopia 105	17.	Iran 82
3.	U.S.A. 327	8.	Bangladesh 158	13.	Philippines 104	18.	Turkey 81
4.	Indonesia 261	9.	Russia 142	14.	Egypt 97	19.	Germany 81
5.	Brazil 207	10.	Japan 126	15.	Vietnam 96	20.	Thailand 68

Equatorial Scale 1:100 000 000
This distance is 4,000 kilometres

	SMALLEST COUNTRIES (thousand people)		
1. Vatican City 1	6. San Marino 34	11. Andorra 86	16. Tonga 106
2. Nauru 10	7. Liechtenstein 38	12. Seychelles 94	17. Kiribati 108
3. Tuvalu 11	8. St. Kitts & Nevis 52	13. Antigua & Barbuda 95	18. Grenada 112
4. Palau 21	9. Dominica 74	14. St. Vincent & The Grenadines ... 102	19. St. Lucia 164
5. Monaco 31	10. Marshall Islands 75	15. Micronesia, Fed. States of 104	20. Samoa 200

COPYRIGHT PHILIP'S

AT A GLANCE

- North America is the third largest continent. It is half the size of Asia. It stretches almost from the Equator to the North Pole.
- Three countries – Canada, the United States and Mexico – make up most of the continent.
- Greenland, the largest island in the world, is part of North America.

- In the east there are several large lakes. These are called the Great Lakes. A large waterfall called Niagara Falls is between Lake Erie and Lake Ontario. The St Lawrence River connects the Great Lakes with the Atlantic Ocean.
- North and South America are joined by the Isthmus of Panama.

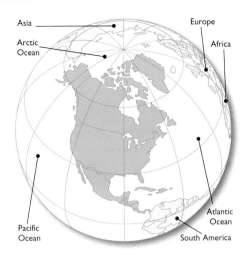

Mount Denali (once named McKinley) is the highest peak in North America.

Lake Ontario, one of the Great Lakes, flows over the Niagara Falls into Lake Erie.

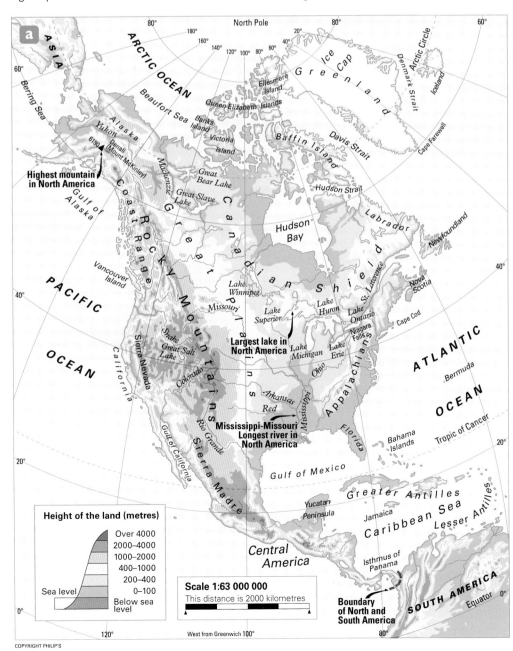

Scale 1:63 000 000
This distance is 2000 kilometres

FACT FILE

LARGEST COUNTRIES:
BY AREA
(thousand square kilometres)

1.	Canada	9,971
2.	United States	9,629
3.	Mexico	1,958
4.	Nicaragua	129
5.	Honduras	112
6.	Cuba	111
7.	Guatemala	109
8.	Panama	76
9.	Costa Rica	51
10.	Dominican Republic	49

LARGEST COUNTRIES:
BY POPULATION
(million people)

1.	United States	327
2.	Mexico	125
3.	Canada	36
4.	Guatemala	15
5.	Cuba	11
6.	Haiti	11
7.	Dominican Republic	10
8.	Honduras	9
9.	El Salvador	6
10.	Nicaragua	6

LARGEST CITIES
(million people)

1.	Mexico City, Mexico	21.2
2.	New York, USA	20.1
3.	Los Angeles, USA	13.3
4.	Chicago, USA	9.6
5.	Dallas-Fort Worth, USA	7.0
6.	Houston, USA	6.5
7.	Toronto, Canada	6.1
8.	Philadelphia, USA	6.1
9.	Washington, DC, USA	6.0
10.	Miami, USA	5.9

COPYRIGHT PHILIP'S

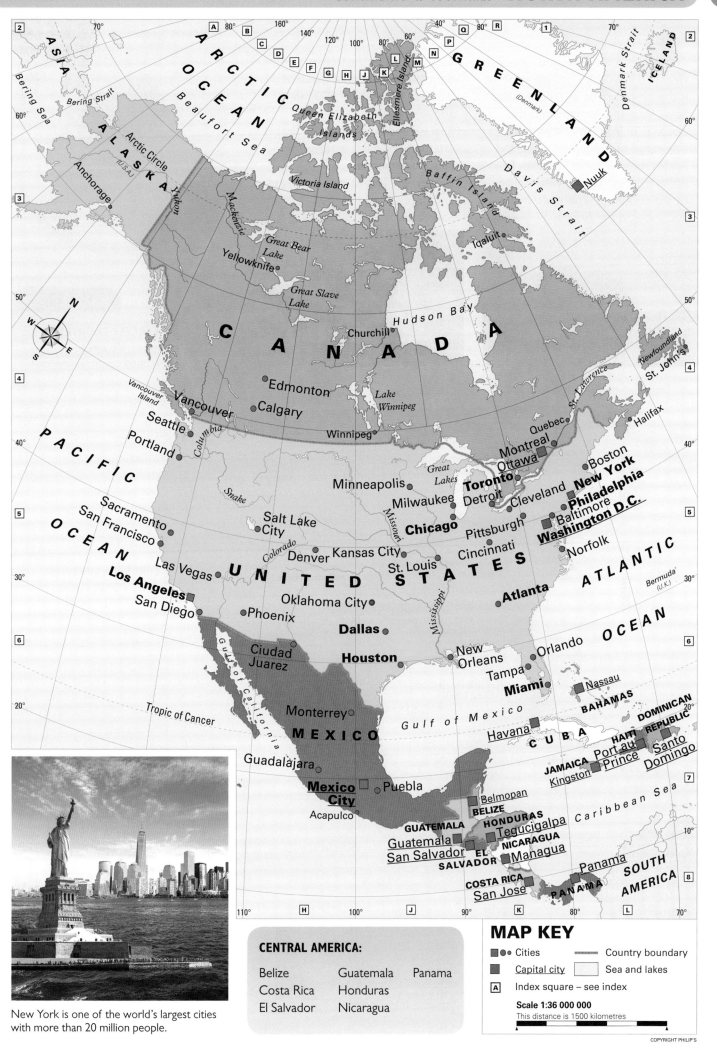

ASIA

ARCTIC OCEAN

Bering Sea
Bering Strait

ALASKA (U.S.A.)
Arctic Circle
Anchorage

Beaufort Sea
Yukon

G R E E N L A N D (Denmark)

Denmark Strait
ICELAND

Queen Elizabeth Islands
Ellesmere Island

Davis Strait

Nuuk

Victoria Island

Baffin Island

Iqaluit

Mackenzie

Great Bear Lake
Yellowknife

Great Slave Lake

C A N A D A

Hudson Bay
Churchill

Lake Winnipeg

St. Lawrence

Newfoundland
St. John's

Edmonton
Vancouver Island
Vancouver
Calgary

Seattle
Portland

Columbia
Snake

Winnipeg

Quebec
Montreal
Ottawa

Halifax

PACIFIC

Sacramento
San Francisco

O C E A N

Las Vegas

Salt Lake City

Minneapolis
Milwaukee
Chicago

Great Lakes
Detroit

Toronto

Cleveland

Boston
New York
Philadelphia
Baltimore
Washington D.C.

Missouri

Pittsburgh

Norfolk

ATLANTIC

Los Angeles
San Diego

Colorado
Denver
Kansas City

Phoenix

U N I T E D   S T A T E S

Oklahoma City

St. Louis

Cincinnati

Atlanta

Bermuda (U.K.)

O C E A N

Dallas

Mississippi

Houston

New Orleans

Orlando
Tampa

Gulf of California
Ciudad Juarez

Tropic of Cancer

Monterrey

M E X I C O

Miami

Nassau

BAHAMAS

Havana

Gulf of Mexico

C U B A

HAITI

DOMINICAN REPUBLIC

Santo Domingo

Guadalajara

Mexico City
Acapulco

Puebla

JAMAICA
Kingston

Port au Prince

Caribbean Sea

Belmopan
BELIZE
GUATEMALA
Guatemala
San Salvador
EL SALVADOR

HONDURAS
Tegucigalpa
NICARAGUA
Managua

Panama

SOUTH AMERICA

COSTA RICA
San Jose

PANAMA

New York is one of the world's largest cities with more than 20 million people.

CENTRAL AMERICA:

Belize   Guatemala   Panama
Costa Rica   Honduras
El Salvador   Nicaragua

MAP KEY

■●● Cities   ——— Country boundary
■ Capital city   Sea and lakes
Ⓐ Index square – see index

Scale 1:36 000 000
This distance is 1500 kilometres

COPYRIGHT PHILIP'S

AT A GLANCE

- The Amazon is the second longest river in the world. The Nile in Africa is the longest river, but more water flows from the Amazon into the ocean than from any other river.
- The range of mountains called the Andes runs for over 7,500 km from north to south on the western side of the continent. There are many volcanoes in the Andes.

- Lake Titicaca is the largest lake in the continent. It has an area of 8,300 sq km and is 3,800 metres above sea level.
- Spanish and Portuguese (in Brazil) are the principal languages spoken in South America.
- Brazil is the largest country in area and population, and the largest city, Sao Paulo.

Snow-covered Andes Mountains tower above Lake Titicaca which is 3812 metres high.

The Amazon river system empties into the south Atlantic Ocean at the Equator.

Scale 1:60 000 000
This distance is 2000 kilometres

COPYRIGHT PHILIP'S

Height of the land (metres)

| Over 4000 |
| 2000–4000 |
| 1000–2000 |
| 400–1000 |
| 200–400 |
| 0–100 |
| Sea level |
| Below sea level |

FACT FILE

LARGEST COUNTRIES: BY AREA
(thousand square kilometres)

1.	Brazil	8,514
2.	Argentina	2,780
3.	Peru	1,285
4.	Colombia	1,139
5.	Bolivia	1,099
6.	Venezuela	912
7.	Chile	757
8.	Paraguay	407
9.	Ecuador	284
10.	Guyana	215

LARGEST COUNTRIES: BY POPULATION
(million people)

1.	Brazil	207
2.	Colombia	48
3.	Argentina	44
4.	Venezuela	31
5.	Peru	31
6.	Chile	18
7.	Ecuador	16
8.	Bolivia	11
9.	Paraguay	7
10.	Uruguay	3

LARGEST CITIES
(million people)

1.	Sao Paulo, Brazil	21.3
2.	Buenos Aires, Argentina	15.3
3.	Rio de Janeiro, Brazil	13.0
4.	Lima, Peru	10.1
5.	Bogota, Colombia	10.0
6.	Santiago, Chile	6.5
7.	Belo Horizonte, Brazil	5.8
8.	Brasilia, Brazil	4.2
9.	Medellin, Colombia	4.0
10.	Fortaleza, Brazil	3.9

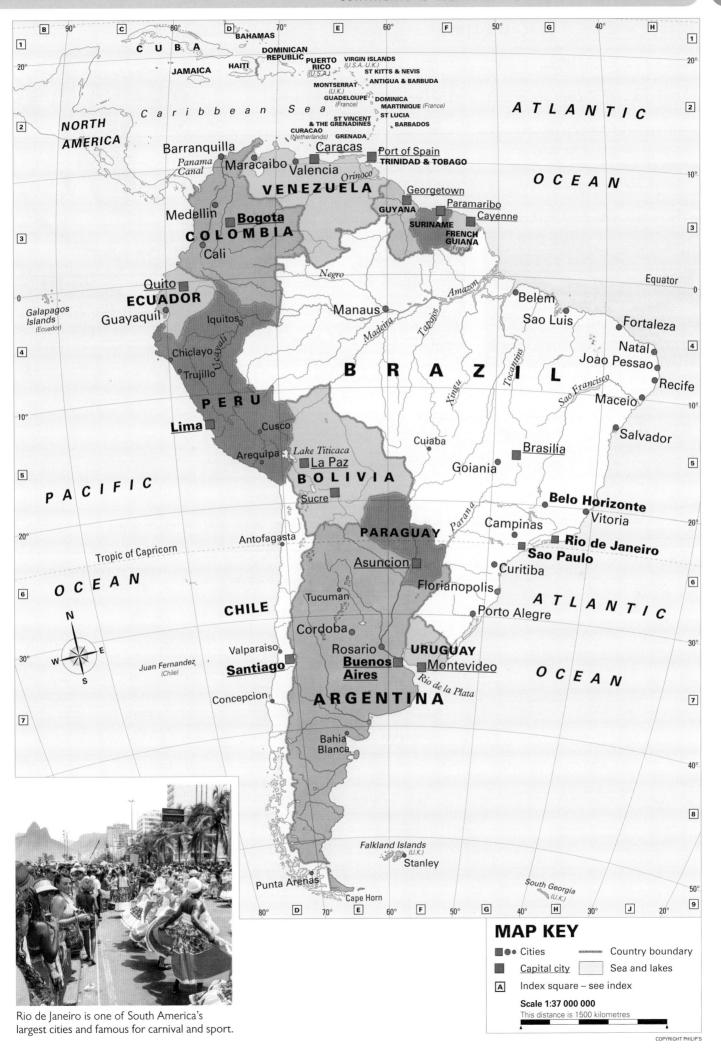

CUBA
BAHAMAS
DOMINICAN
REPUBLIC
JAMAICA   HAITI   PUERTO   VIRGIN ISLANDS
                      RICO   (U.S.A.-U.K.)
                      (U.S.A.) ST KITTS & NEVIS
                                ANTIGUA & BARBUDA
                      MONTSERRAT
                      (U.K.)
                      GUADELOUPE   DOMINICA
                      (France)   MARTINIQUE (France)
                      ST VINCENT   ST LUCIA
                      & THE GRENADINES
              CURACAO   BARBADOS
              (Netherlands)   GRENADA

NORTH
AMERICA

Caribbean Sea

*Panama
Canal*

ATLANTIC

OCEAN

Barranquilla   Caracas
Maracaibo   Port of Spain
Valencia   TRINIDAD & TOBAGO
              Orinoco
          VENEZUELA
                          Georgetown
Medellin   **Bogota**   Paramaribo
                      GUYANA   Cayenne
COLOMBIA   SURINAME FRENCH
   Cali   GUIANA
                               (France)

Negro   Equator

Quito   *Amazon*
ECUADOR   Belem
Guayaquil   Sao Luis
          Iquitos   Fortaleza
*Galapagos   *Madeira* *Tapajós* Natal
Islands*   Chiclayo Manaus   Joao Pessao
(Ecuador) *Ucayali*   Recife
   Trujillo   **B R A Z I L**   Maceio
                    Xingu *Tocantins* *Sao Francisco*
PERU   Salvador
Lima   Cusco   Cuiaba
        Arequipa   *Lake Titicaca*   **Brasilia**
              La Paz   Goiania
              BOLIVIA
              Sucre   **Belo Horizonte**
                                     Vitoria
                 Parana   Campinas   **Rio de Janeiro**
           PARAGUAY   **Sao Paulo**
Antofagasta   Curitiba
        Asuncion   Florianopolis
Tropic of Capricorn

PACIFIC   Tucuman
              CHILE   Porto Alegre
OCEAN   Cordoba
        Valparaiso   Rosario   **URUGUAY**   ATLANTIC
   Juan Fernandez   **Buenos** **Montevideo**
   (Chile) **Santiago** **Aires** *Rio de la Plata* OCEAN
        Concepcion   **ARGENTINA**

N
W   E
S

Bahia
Blanca

Falkland Islands
(U.K.)
   Stanley

Punta Arenas
Cape Horn   *South Georgia*
                  (U.K.)

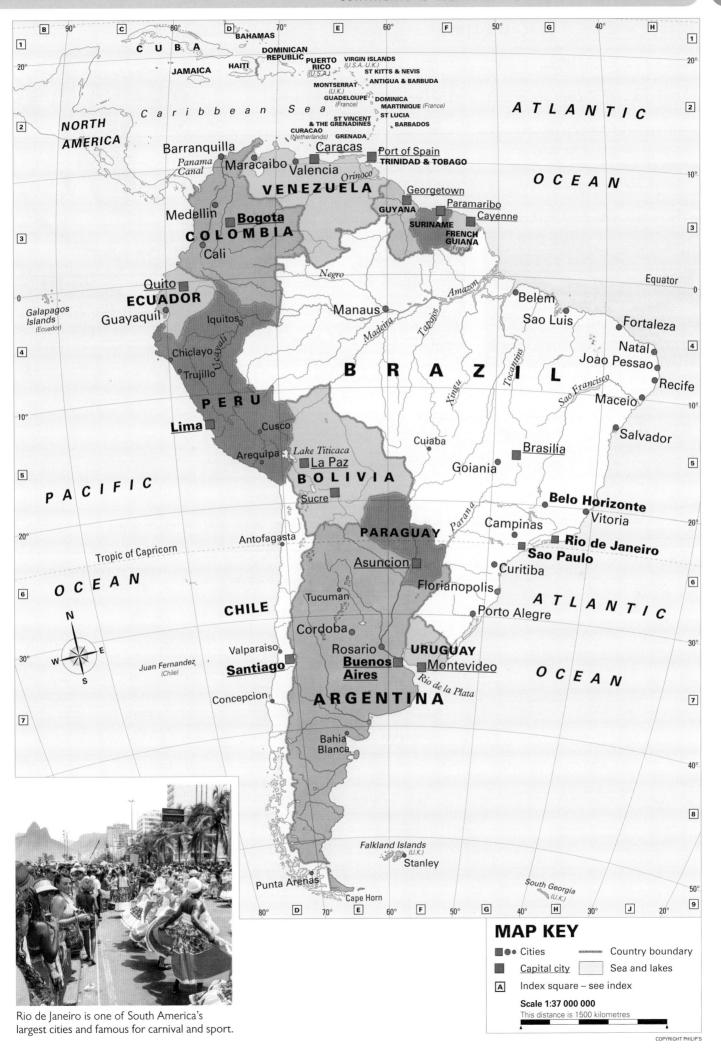
Rio de Janeiro is one of South America's
largest cities and famous for carnival and sport.

MAP KEY

■ ● Cities   ‑‑‑‑‑ Country boundary

■ Capital city   ☐ Sea and lakes

Ⓐ Index square – see index

Scale 1:37 000 000
This distance is 1500 kilometres

COPYRIGHT PHILIP'S

AT A GLANCE

- Africa is the second largest continent. Asia is the largest.
- There are over 50 countries, some of them small in population. The population of Africa is growing more quickly than any other continent.
- Parts of Africa have a dry, desert climate. Some other parts are tropical.
- The highest mountains run from north to south on the eastern side of Africa.
- The Great Rift Valley is a volcanic valley that was formed 10 to 20 million years ago through long splits in the Earth's crust. Mount Kenya and Kilimanjaro are old volcanoes in the Rift Valley area.
- The Sahara is the largest desert in the world.

Sand dunes of the Sahara result from the dry, desert climate of much of North Africa

Lake Tanganyika is in a section of the Great Rift Valley. The entire valley is 6000 km long.

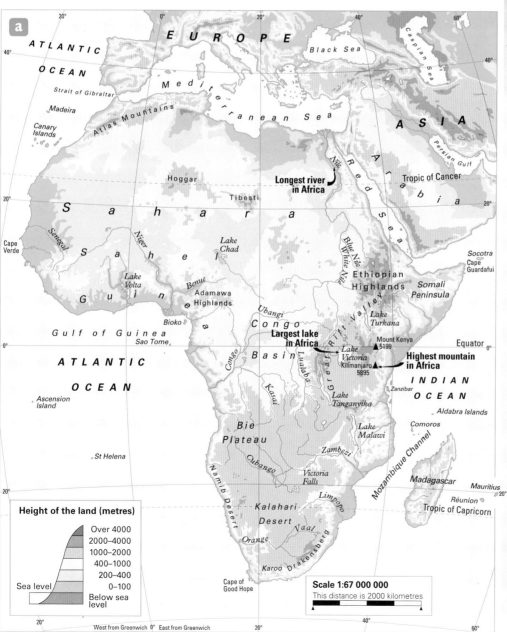

FACT FILE

LARGEST COUNTRIES: BY AREA
(thousand square kilometres)

1. Algeria 2,382
2. Dem. Rep. of the Congo 2,345
3. Sudan 1,886
4. Libya 1,759
5. Chad 1,284
6. Niger 1,267
7. Angola 1,247
8. Mali 1,240
9. South Africa 1,221
10. Ethiopia 1,104

LARGEST COUNTRIES: BY POPULATION
(million people)

1. Nigeria 191
2. Ethiopia 105
3. Egypt 97
4. Dem. Rep. of the Congo . 83
5. South Africa 55
6. Tanzania 54
7. Kenya 48
8. Algeria 41
9. Uganda 40
10. Sudan 37

LARGEST CITIES
(million people)

1. Cairo, Egypt 17.1
2. Lagos, Nigeria 13.7
3. Kinshasa, Dem. Rep. of the Congo 12.1
4. Johannesburg, S. Africa . . 9.6
5. Luanda, Angola 5.7
6. Dar es Salaam, Tanzania . . 5.4
7. Khartoum, Sudan 5.2
8. Abidjan, Côte d'Ivoire . . . 5.0
9. Alexandria, Egypt 4.9
10. Nairobi, Kenya 4.0

COPYRIGHT PHILIP'S

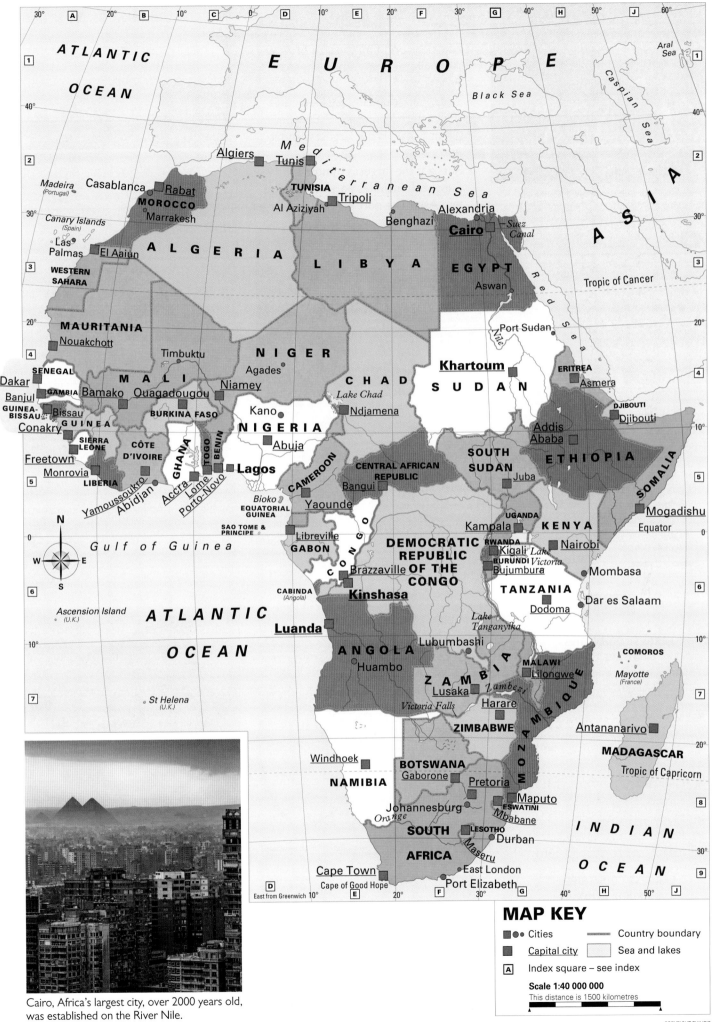

ATLANTIC OCEAN

EUROPE

ASIA

Black Sea

Aral Sea

Caspian Sea

Mediterranean Sea

Algiers
Tunis
TUNISIA
Casablanca
Madeira (Portugal)
Rabat
MOROCCO
Marrakesh
Tripoli
Al Aziziyah
Benghazi
Alexandria
Cairo
Suez Canal
Canary Islands (Spain)
Las Palmas
El Aaiun
WESTERN SAHARA
ALGERIA
LIBYA
EGYPT
Aswan
Red Sea
Tropic of Cancer

MAURITANIA
Nouakchott
Timbuktu
NIGER
Agades
Port Sudan
Nile
ERITREA
Asmera
SENEGAL
MALI
Niamey
CHAD
Khartoum
SUDAN
Dakar
Banjul
GAMBIA
Bamako
Ouagadougou
Lake Chad
Ndjamena
DJIBOUTI
Djibouti
GUINEA-BISSAU
Bissau
BURKINA FASO
Kano
NIGERIA
Addis Ababa
ETHIOPIA
Conakry
GUINEA
SIERRA LEONE
CÔTE D'IVOIRE
GHANA
TOGO
BENIN
Abuja
SOUTH SUDAN
Freetown
Monrovia
LIBERIA
Accra
Lome
Porto-Novo
Lagos
CAMEROON
CENTRAL AFRICAN REPUBLIC
Bangui
Juba
SOMALIA
Yamoussoukro
Abidjan
Bioko
EQUATORIAL GUINEA
Yaounde
Mogadishu
Equator

N
W E
S

Gulf of Guinea

SAO TOME & PRINCIPE
Libreville
GABON
CONGO
UGANDA
Kampala
RWANDA
Kigali
BURUNDI
Bujumbura
Lake Victoria
KENYA
Nairobi
Mombasa

DEMOCRATIC REPUBLIC OF THE CONGO
Brazzaville
Kinshasa

Ascension Island (U.K.)

ATLANTIC OCEAN

CABINDA (Angola)
Luanda

TANZANIA
Dodoma
Dar es Salaam

Lake Tanganyika

St Helena (U.K.)

ANGOLA
Huambo
Lubumbashi

COMOROS
Mayotte (France)

MALAWI
Lilongwe

ZAMBIA
Lusaka
Zambezi
Victoria Falls
Harare
ZIMBABWE

MOZAMBIQUE
Antananarivo
MADAGASCAR
Tropic of Capricorn

Windhoek
BOTSWANA
Gaborone
NAMIBIA
Pretoria
Maputo
ESWATINI
Mbabane
INDIAN OCEAN

Johannesburg
Orange
SOUTH AFRICA
LESOTHO
Maseru
Durban

Cape Town
Cape of Good Hope
East London
Port Elizabeth

East from Greenwich 10°

OCEAN

MAP KEY

■●● Cities

■ Capital city

Ⓐ Index square – see index

⸺ Country boundary

▢ Sea and lakes

Scale 1:40 000 000
This distance is 1500 kilometres

Cairo, Africa's largest city, over 2000 years old, was established on the River Nile.

The Ural Mountain range divides one land mass into the continents of Europe and Asia.

AT A GLANCE

- Europe is one-fifth the size of Asia. Australia is slightly smaller than Europe.
- The Ural Mountains are viewed as the eastern boundary of Europe.
- Great Britain is the largest island in Europe
- Russia is the largest country in Europe. It includes parts of Europe and Asia. The part in Asia is far larger.

FACT FILE

LARGEST COUNTRIES: BY AREA
(thousand square kilometres)

1. Russia (in Europe) 3,960
2. Ukraine 604
3. France 551
4. Spain 497
5. Sweden 450
6. Germany 357
7. Finland 338
8. Norway 324
9. Poland 323
10. Italy 301

LARGEST COUNTRIES: BY POPULATION
(million people)

1. Russia (in Europe) 110
2. Germany 81
3. France 67
4. United Kingdom 65
5. Italy 62

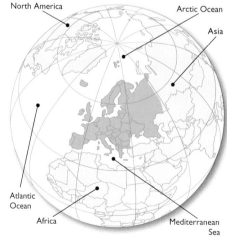

The Highlands of Scotland are largely unihabited, but contain some of the finest scenery in the British Isles, including Ben Nevis, its highest mountain.

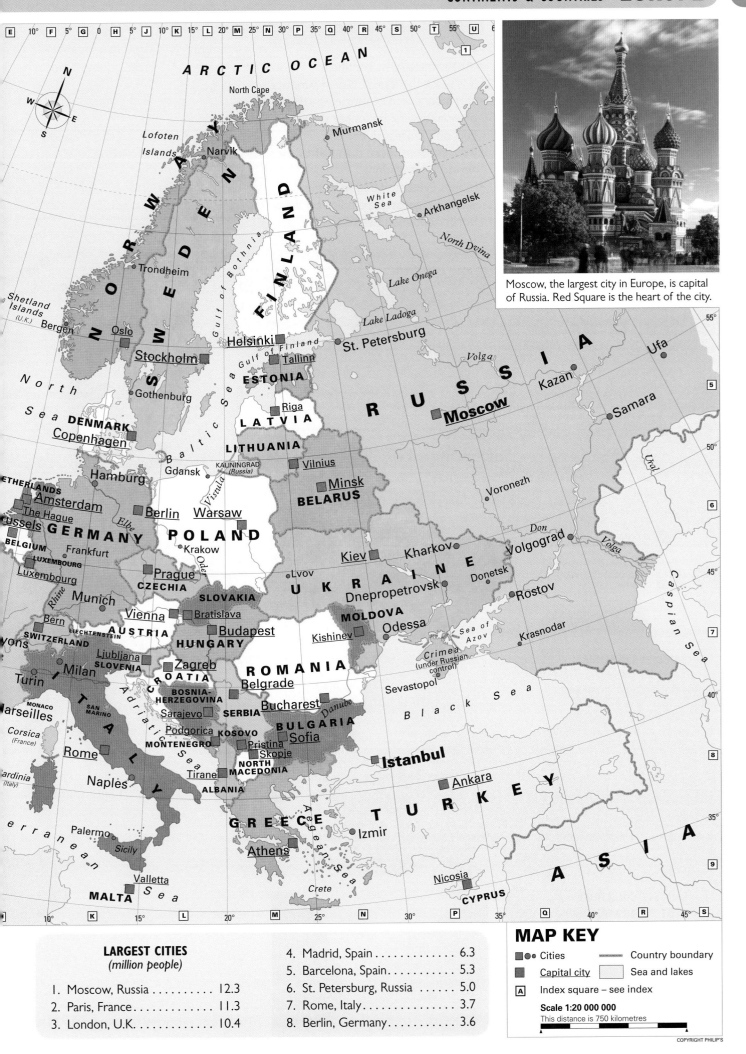

Moscow, the largest city in Europe, is capital of Russia. Red Square is the heart of the city.

LARGEST CITIES
(million people)

1. Moscow, Russia 12.3
2. Paris, France 11.3
3. London, U.K. 10.4
4. Madrid, Spain 6.3
5. Barcelona, Spain 5.3
6. St. Petersburg, Russia 5.0
7. Rome, Italy 3.7
8. Berlin, Germany 3.6

MAP KEY

■●● Cities	- - - - Country boundary
■ Capital city	Sea and lakes
Ⓐ Index square – see index	

Scale 1:20 000 000
This distance is 750 kilometres

COPYRIGHT PHILIP'S

Mount Everest viewed from Nepal.

AT A GLANCE

- Asia is the largest continent. It is twice the size of North America.
- It is a continent of long rivers. Many of Asia's rivers are longer than Europe's longest river.
- Asia contains more than half of the world's population.
- Mount Everest, 8,850 m, is the world's highest peak. It lies on the border between Nepal and China.

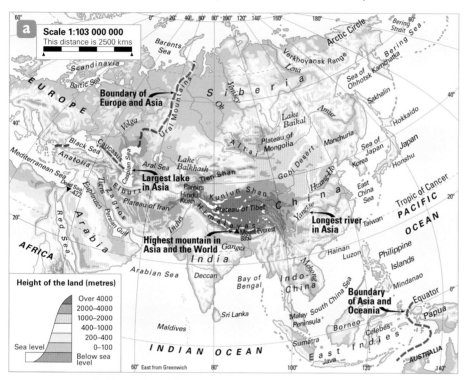

Scale 1:103 000 000
This distance is 2500 kms

Height of the land (metres)
Over 4000
2000–4000
1000–2000
400–1000
200–400
0–100
Sea level
Below sea level

60° East from Greenwich   80°   100°   120°   140°

FACT FILE

LARGEST COUNTRIES: BY AREA
(thousand square kilometres)

1. Russia (in Asia) 13,115
2. China 9,597
3. India 3,287
4. Kazakhstan 2,725
5. Saudi Arabia 2,150
6. Indonesia 1,905
7. Iran 1,648
8. Mongolia 1,566
9. Pakistan 796
10. Turkey 775

LARGEST COUNTRIES: BY POPULATION
(million people)

1. China 1,379
2. India 1,282
3. Indonesia 261
4. Pakistan 205
5. Bangladesh 158

Arctic Ocean
North America
Pacific Ocean
Europe
Africa
Indian Ocean
Oceania

The Yangtse river winds 6000 km across China from west to east.

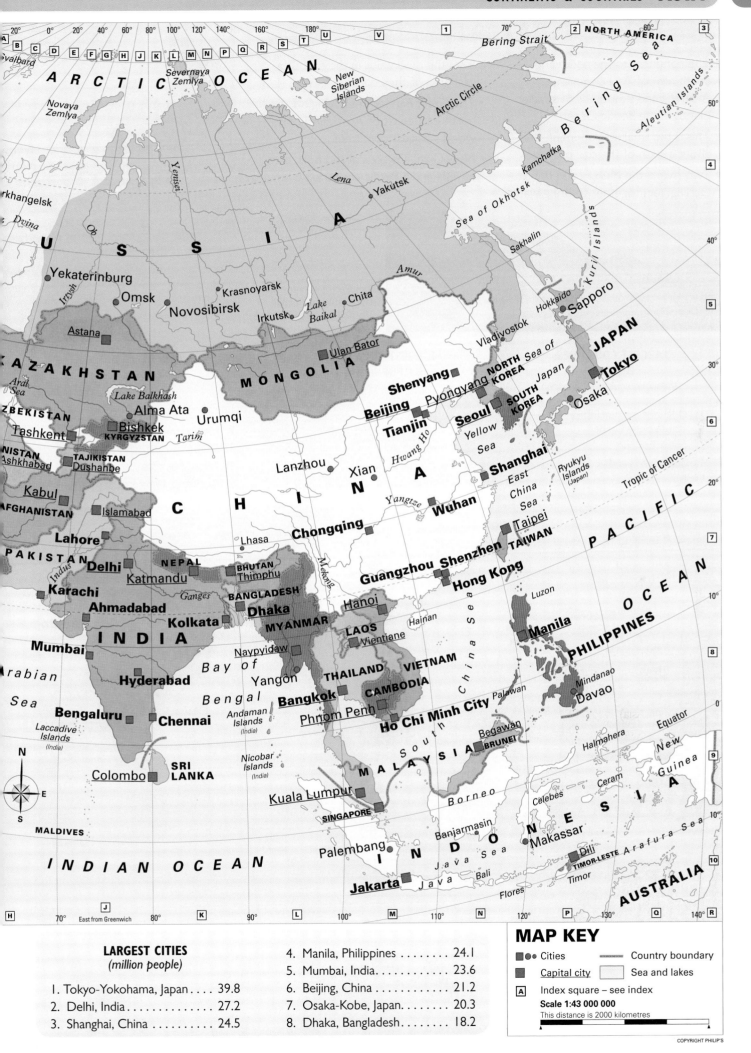

LARGEST CITIES
(million people)

1. Tokyo-Yokohama, Japan 39.8
2. Delhi, India 27.2
3. Shanghai, China 24.5
4. Manila, Philippines 24.1
5. Mumbai, India. 23.6
6. Beijing, China 21.2
7. Osaka-Kobe, Japan. 20.3
8. Dhaka, Bangladesh. 18.2

MAP KEY

■●● Cities
■ Capital city
Ⓐ Index square – see index

—·—·— Country boundary
Sea and lakes

Scale 1:43 000 000
This distance is 2000 kilometres

COPYRIGHT PHILIP'S

AT A GLANCE

- The continent to the south and south-east of Asia comprises Australia and thousands of smaller islands in the Pacific Ocean

- This is the smallest continent, only about a sixth of the size of Asia.

- Two rivers, the Murray and the Darling, join together to bring water from the mountains in the east to a vast area of the country.

- The Great Barrier Reef, lying off the north-east coast of Australia, is the world's biggest coral reef.

- New Guinea is considered both part of Asia and a Pacific country.

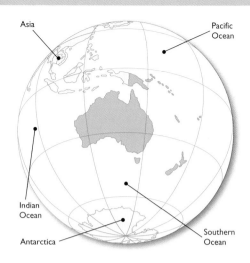

Many Pacific countries have islands which are active volcanoes, such as this one in Vanuatu.

LARGEST COUNTRIES: BY AREA (thousand square kilometres)	**LARGEST COUNTRIES: BY POPULATION** (million people)	**LARGEST CITIES** (million people)
1. Australia 7,741	1. Australia 23	1. Sydney, Australia 4.5
2. Papua New Guinea 463	2. Papua New Guinea 7	2. Melbourne, Australia 4.3
3. New Zealand 270	3. New Zealand 5	3. Brisbane, Australia 2.2
4. Solomon Islands 26	4. Fiji 0.9	4. Perth, Australia 1.9
5. Fiji 18	5. Solomon Islands 0.6	5. Auckland, New Zealand 1.4

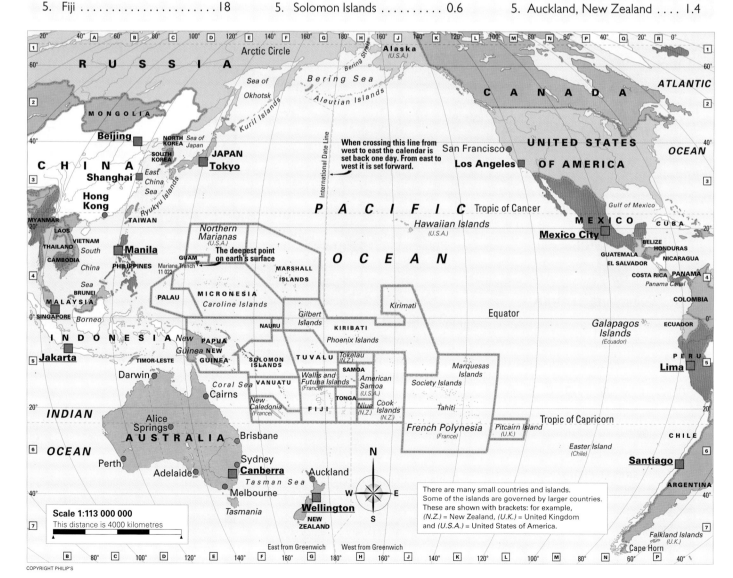

When crossing this line from west to east the calendar is set back one day. From east to west it is set forward.

The deepest point on earth's surface

Mariana Trench 11 022

There are many small countries and islands. Some of the islands are governed by larger countries. These are shown with brackets: for example, (N.Z.) = New Zealand, (U.K.) = United Kingdom and (U.S.A.) = United States of America.

Scale 1:113 000 000
This distance is 4000 kilometres

NORTH POLE a

PACIFIC OCEAN

Anchorage

ARCTIC OCEAN

Nizhne Kolymsk

Wrangel I. (Russia)

Tiksi

New Siberian Is. (Russia)

Laptev Sea

ASIA

Novosibirsk

Beaufort Sea

Taimyr Peninsula

Norilsk

C. Chelyuskin

Severnaya Zemlya (Russia)

Inuvik

North + Magnetic Pole

North Pole

Franz Josef Land (Russia)

Kara Sea

Vorkuta

Ural Mts.

Vancouver

Banks Island (Canada)

Queen Elizabeth Islands (Canada)

Novaya Zemlya (Russia)

Great Bear L.

Yellowknife

Victoria Island (Canada)

Great Slave L.

Barents Sea

Edmonton

C. Morris Jesup

Ellesmere Island (Canada)

Arkhangelsk

NORTH AMERICA

Thule

Baffin Bay

Greenland

Svalbard (Norway)

North Cape

Murmansk

Greenland Sea

Tromsø

Nelson

Churchill

Baffin Island (Canada)

Moscow

Hudson Bay

Hudson Strait

Davis Strait

Jan Mayen I. (Norway)

St. Petersburg

Iqaluit

Arctic Circle

Oslo

L. Michigan

L. Superior

Labrador

Nuuk (Godthåb)

Denmark Strait

Iceland

Reykjavik

Faroe Is. (Denmark)

Baltic Sea

Black Sea

Chicago

L. Huron

C. Farewell

North Sea

British Isles

Edinburgh

EUROPE

Scale 1:50 000 000
This distance is 2500 kilometres

West from Greenwich 0° East from Greenwich

Height of the land (metres)

Over 4000
2000–4000
1000–2000
400–1000
200–400
0–200
Below sea level

Sea level

Cities

Capital city

Index square – see index

Davis (Austr.) Research station and the country which runs it.

Limit of permanently frozen sea

Icebergs

Furthest extent of icebergs

Land permanently covered with ice

Height of ice (in metres)

SOUTH POLE b

ATLANTIC OCEAN

South Sandwich Trench ▼8265

South Sandwich Is. (U.K.)

Syowa (Japan)

Enderby Land

Mawson (Austr.)

INDIAN OCEAN

South Georgia (U.K.)

Maitri (India)

C. Darnley

Sanae (S. Africa)

Prince Charles Mts.

Davis (Austr.)

Queen Maud Land

American Highland

South Orkney Is.

Halley (U.K.)

Coats Land

Scotia Sea

Weddell Sea

ANTARCTICA

O'Higgins (Chile)

Casey (Austr.)

Falkland Is. (U.K.)

South Shetland Is.

Esperanza (Arg.)

Berkner I.

Ronne Ice Shelf

Vostok (Russia)

Wilkes Land

Palmer (U.S.A.)

Antarctic Peninsula

South Pole

Amundsen-Scott (U.S.A.)

Rothera (U.K.)

Queen Maud Ra.

Beardmore Glacier

Mt. Markham ▲ 4349

South Magnetic + Pole

Drake Passage

C. Horn

Alexander I.

Vinson Massif 5140

Ellsworth Land

Ross Ice Shelf

Adélie Land

Dumont d'Urville (France)

Tierra del Fuego

Bellingshausen Sea

McMurdo (U.S.A.)

Punta Arenas

Strait of Magellan

Marie Byrd Land

Scott (N.Z.)

Victoria Land

SOUTH AMERICA

Ross Sea

C. Adare

Balleny Is.

Tasmania

Hobart

SOUTHERN OCEAN

Macquarie I. (Australia)

AUSTRALIA

Scale 1:50 000 000
This distance is 2500 kilometres

Antarctic Circle

Campbell I. (N.Z.)

Auckland I.

West from Greenwich 180° East from Greenwich

COPYRIGHT PHILIP'S

Country	Flag	Capital	Area (sq km)	Population (thousands)
NORTH AMERICA				
ANTIGUA & BARBUDA		St John's	442	102
BAHAMAS, THE		Nassau	13,878	351
BARBADOS		Bridgetown	431	290
BELIZE		Belmopan	22,966	388
CANADA		Ottawa	9,970,610	35,624
COSTA RICA		San José	51,100	4,930
CUBA		Havana	110,861	11,239
DOMINICA		Roseau	751	74
DOMINICAN REPUBLIC		Santo Domingo	48,511	10,169
EL SALVADOR		San Salvador	21,041	6,172
GRENADA		St George's	344	108
GUATEMALA		Guatemala City	108,889	15,461
HAITI		Port-au-Prince	27,750	10,912
HONDURAS		Tegucigalpa	112,088	9,039
JAMAICA		Kingston	10,991	2,698
MEXICO		Mexico City	1,958,201	124,575
NICARAGUA		Managua	129,494	6,026
PANAMA		Panamá	75,517	3,753
ST KITTS & NEVIS		Basseterre	261	53
ST LUCIA		Castries	616	180
ST VINCENT & THE GRENADINES		Kingstown	388	110
UNITED STATES OF AMERICA		Washington, DC	9,629,091	326,626
SOUTH AMERICA				
ARGENTINA		Buenos Aires	2,780,400	44,293
BOLIVIA		La Paz/Sucre	1,098,581	11,138
BRAZIL		Brasília	8,514,215	207,353
CHILE		Santiago	756,626	17,789
COLOMBIA		Bogotá	1,138,914	47,699
ECUADOR		Quito	283,561	16,291
FRENCH GUIANA		Cayenne	90,000	250
GUYANA		Georgetown	216,970	800
PARAGUAY		Asunción	406,752	6,944
PERU		Lima	1,285,216	31,037

Country	Flag	Capital	Area (sq km)	Population (thousands)
SURINAME		Paramaribo	163,265	592
TRINIDAD & TOBAGO		Port of Spain	5,128	1,380
URUGUAY		Montevideo	175,016	3,360
VENEZUELA		Caracas	912,050	31,304
AFRICA				
ALGERIA		Algiers	2,381,741	40,969
ANGOLA		Luanda	1,246,700	29,310
BENIN		Porto-Novo	112,622	11,039
BOTSWANA		Gaborone	581,730	2,215
BURKINA FASO		Ouagadougou	274,200	20,108
BURUNDI		Bujumbura	27,834	11,467
CABO VERDE		Praia	4,033	561
CAMEROON		Yaoundé	475,442	24,995
CENTRAL AFRICAN REPUBLIC		Bangui	622,984	5,625
CHAD		Ndjamena	1,284,000	12,076
COMOROS		Moroni	2,235	808
CONGO		Brazzaville	342,000	4,955
CÔTE D'IVOIRE		Yamoussoukro	322,463	24,185
DEMOCRATIC REPUBLIC OF THE CONGO		Kinshasa	2,344,858	83,301
DJIBOUTI		Djibouti	23,200	865
EGYPT		Cairo	1,001,449	97,041
EQUATORIAL GUINEA		Malabo	28,051	778
ERITREA		Asmara	117,600	5,919
ESWATINI		Mbabane	17,364	1,467
ETHIOPIA		Addis Ababa	1,104,300	105,350
GABON		Libreville	267,668	1,772
GAMBIA, THE		Banjul	11,295	2,051
GHANA		Accra	238,533	27,500
GUINEA		Conakry	245,857	12,414
GUINEA-BISSAU		Bissau	36,125	1,792
KENYA		Nairobi	580,367	47,616
LESOTHO		Maseru	30,355	1,958
LIBERIA		Monrovia	111,369	4,689
LIBYA		Tripoli	1,759,540	6,653

The population figures are 2018 estimates where available

Country	Flag	Capital	Area (sq km)	Population (thousands)
MADAGASCAR		Antananarivo	587,041	25,054
MALAWI		Lilongwe	118,484	19,196
MALI		Bamako	1,240,192	17,885
MAURITANIA		Nouakchott	1,025,520	3,759
MAURITIUS		Port Louis	2,040	1,356
MOROCCO		Rabat	446,550	33,987
MOZAMBIQUE		Maputo	801,590	26,574
NAMIBIA		Windhoek	824,292	2,485
NIGER		Niamey	1,267,000	19,245
NIGERIA		Abuja	923,768	190,632
RWANDA		Kigali	26,338	11,901
SÃO TOMÉ & PRÍNCIPE		São Tomé	964	201
SENEGAL		Dakar	196,722	14,669
SEYCHELLES		Victoria	455	94
SIERRA LEONE		Freetown	71,740	6,163
SOMALIA		Mogadishu	637,657	11,031
SOUTH AFRICA		Cape Town/Pretoria	1,221,037	54,842
SOUTH SUDAN		Juba	620,000	13,026
SUDAN		Khartoum	1,886,086	37,346
TANZANIA		Dodoma	945,090	53,951
TOGO		Lomé	56,785	7,965
TUNISIA		Tunis	163,610	11,404
UGANDA		Kampala	241,038	39,570
WESTERN SAHARA		El Aaiún	266,000	603
ZAMBIA		Lusaka	752,618	15,972
ZIMBABWE		Harare	390,757	13,805

EUROPE

Country	Flag	Capital	Area (sq km)	Population (thousands)
ALBANIA		Tirana	28,748	3,048
ANDORRA		Andorra La Vella	468	86
AUSTRIA		Vienna	83,859	8,754
BELARUS		Minsk	207,600	9,550
BELGIUM		Brussels	30,528	11,491
BOSNIA-HERZEGOVINA		Sarajevo	51,197	3,856
BULGARIA		Sofia	110,912	7,102

Country	Flag	Capital	Area (sq km)	Population (thousands)
CROATIA		Zagreb	56,538	4,292
CZECHIA		Prague	78,866	10,675
DENMARK		Copenhagen	43,094	5,606
ESTONIA		Tallinn	45,100	1,252
FINLAND		Helsinki	338,145	5,518
FRANCE		Paris	551,500	67,106
GERMANY		Berlin	357,022	80,594
GREECE		Athens	131,957	10,768
HUNGARY		Budapest	93,032	9,851
ICELAND		Reykjavik	103,000	340
IRELAND		Dublin	70,273	5,011
ITALY		Rome	301,318	62,138
KOSOVO		Pristina	10,887	1,895
LATVIA		Riga	64,600	1,945
LIECHTENSTEIN		Vaduz	160	38
LITHUANIA		Vilnius	65,200	2,824
LUXEMBOURG		Luxembourg	2,586	594
MALTA		Valletta	316	416
MOLDOVA		Kishinev	33,851	3,474
MONACO		Monaco	1	31
MONTENEGRO		Podgorica	14,026	643
NETHERLANDS		The Hague	41,526	17,085
NORTH MACEDONIA		Skopje	25,713	2,104
NORWAY		Oslo	323,877	5,320
POLAND		Warsaw	323,250	38,476
PORTUGAL		Lisbon	88,797	10,840
ROMANIA		Bucharest	238,391	21,530
RUSSIA (Europe & Asia)		Moscow	17,075,400	142,258
SAN MARINO		San Marino	61	34
SERBIA		Belgrade	77,474	7,111
SLOVAKIA		Bratislava	49,012	5,446
SLOVENIA		Ljubljana	20,256	1,972
SPAIN		Madrid	497,548	48,958
SWEDEN		Stockholm	449,964	9,961

Country	Flag	Capital	Area (sq km)	Population (thousands)
SWITZERLAND		Bern	41,284	8,236
UKRAINE		Kiev	603,700	44,034
UNITED KINGDOM		London	241,857	64,769
VATICAN CITY		Vatican City	0.44	1

ASIA				
AFGHANISTAN		Kabul	652,090	34,125
ARMENIA		Yerevan	29,800	3,045
AZERBAIJAN		Baku	86,600	9,961
BAHRAIN		Manama	694	1,411
BANGLADESH		Dhaka	143,998	157,827
BHUTAN		Thimphu	47,000	758
BRUNEI		Bandar Seri Begawan	5,765	444
CAMBODIA		Phnom Penh	181,035	16,204
CHINA		Beijing	9,596,961	1,379,303
CYPRUS		Nicosia	9,251	1,222
EAST TIMOR		Dili	14,874	1,291
GEORGIA		Tbilisi	69,700	4,926
INDIA		New Delhi	3,287,263	1,281,936
INDONESIA		Jakarta	1,904,569	260,581
IRAN		Tehran	1,648,195	82,022
IRAQ		Baghdad	438,317	39,192
ISRAEL		Jerusalem	20,600	8,300
JAPAN		Tokyo	377,829	126,451
JORDAN		Amman	89,342	10,248
KAZAKHSTAN		Astana	2,724,900	18,557
KUWAIT		Kuwait City	17,818	2,875
KYRGYZSTAN		Bishkek	199,900	5,789
LAOS		Vientiane	236,800	7,127
LEBANON		Beirut	10,400	6,230
MALAYSIA		Kuala Lumpur	329,758	31,382
MALDIVES		Malé	298	393
MONGOLIA		Ulan Bator	1,566,500	3,068
MYANMAR		Naypyidaw	676,578	55,124
NEPAL		Katmandu	147,181	29,384

Country	Flag	Capital	Area (sq km)	Population (thousands)
NORTH KOREA		Pyŏngyang	120,538	25,248
OMAN		Muscat	309,500	3,424
PAKISTAN		Islamabad	796,095	204,925
PHILIPPINES		Manila	300,000	104,256
QATAR		Doha	11,437	2,314
SAUDI ARABIA		Riyadh	2,149,690	28,572
SINGAPORE		Singapore City	683	5,889
SOUTH KOREA		Seoul	99,268	51,181
SRI LANKA		Colombo	65,610	22,409
SYRIA		Damascus	185,180	18,029
TAIWAN		Taipei	35,980	23,508
TAJIKISTAN		Dushanbe	143,100	8,469
THAILAND		Bangkok	513,115	68,414
TURKEY		Ankara	774,815	80,845
TURKMENISTAN		Ashkhabad	488,100	5,351
UNITED ARAB EMIRATES		Abu Dhabi	83,600	5,927
UZBEKISTAN		Tashkent	447,400	29,749
VIETNAM		Hanoi	331,689	96,160
YEMEN		Sana'	527,968	28,037

AUSTRALIA & THE PACIFIC				
AUSTRALIA		Canberra	7,741,220	23,232
FIJI		Suva	18,274	921
KIRIBATI		Tarawa	726	108
MARSHALL ISLANDS		Majuro	181	75
MICRONESIA		Palikir	702	104
NAURU		Yaren	21	10
NEW ZEALAND		Wellington	270,534	4,510
PALAU		Melekeok	459	21
PAPUA NEW GUINEA		Port Moresby	462,840	6,910
SAMOA		Apia	2,831	200
SOLOMON ISLANDS		Honiara	28,896	648
TONGA		Nuku'alofa	650	106
TUVALU		Fongafale	30	11
VANUATU		Port-Vila	12,189	283

WHERE PEOPLE LIVE

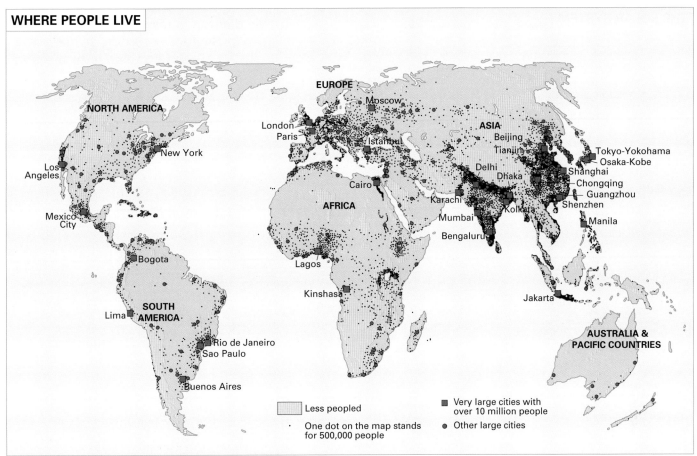

Less peopled

One dot on the map stands for 500,000 people

■ Very large cities with over 10 million people

● Other large cities

Greater Tokyo (the Tokyo-Yokohama metropolitan area) is the largest urban area on Earth.

THE WORLD'S LARGEST CITIES

CITY	CONTINENT	COUNTRY	POPULATION
Tokyo-Yokohama	Asia	Japan	39,800,000
Delhi	Asia	India	27,200,000
Shanghai	Asia	China	24,500,000
Manila	Asia	Philippines	24,100,000
Mumbai	Asia	India	23,600,000
São Paulo	South America	Brazil	21,900,000
Mexico City	North America	Mexico	21,200,000
Beijing	Asia	China	21,200,000

THE GROWTH OF THE POPULATION OF THE WORLD 1750–2018

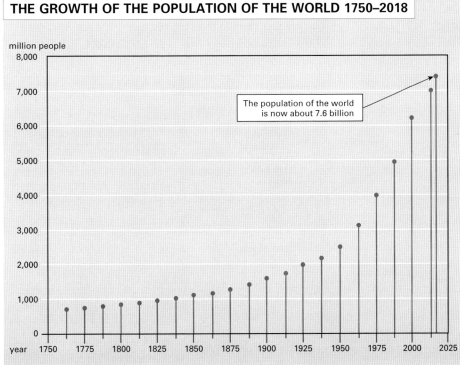

The population of the world is now about 7.6 billion

THE POPULATION OF THE CONTINENTS (2018)

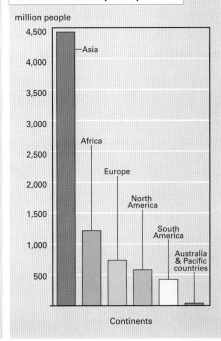

COPYRIGHT PHILIP'S

COPYRIGHT PHILIP'S

THE SOLAR SYSTEM

This diagram shows the planets of the Solar System according to sizes and position relative to the Sun

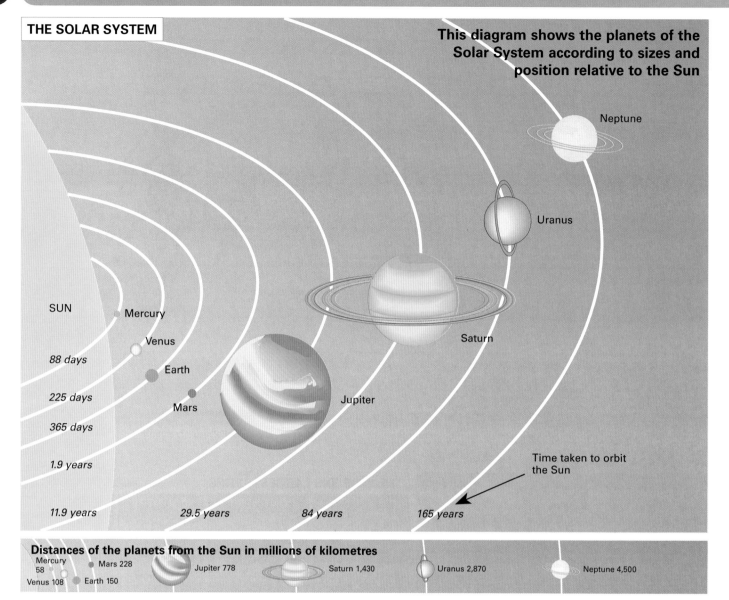

SUN
Mercury
Venus
Earth
Mars
Jupiter
Saturn
Uranus
Neptune

88 days
225 days
365 days
1.9 years
11.9 years   29.5 years   84 years   165 years

Time taken to orbit the Sun

Distances of the planets from the Sun in millions of kilometres

Mercury 58
Venus 108   Earth 150   Mars 228   Jupiter 778   Saturn 1,430   Uranus 2,870   Neptune 4,500

- The Universe is made of many galaxies, or collections of stars. Earth's galaxy is called the Milky Way. It is made up of about 100,000 million stars. The Sun is one of these stars.
- Around the Sun revolve eight planets, one of which is the Earth. The Earth is the fifth largest planet. The Sun, its planets and their satellites are known as the Solar System.
- The Sun is the only source of light and heat in the Solar System. Other planets are visible from the Earth because of the sunlight which they reflect.
- The planets orbit the Sun in the same direction – anti-clockwise when viewed from the northern hemisphere. They also rotate on their own axes.
- The planets remain in orbit because they are attracted by the Sun's pull of gravity. The Earth takes 365 days (a year) to go round the Sun.

THE MOON

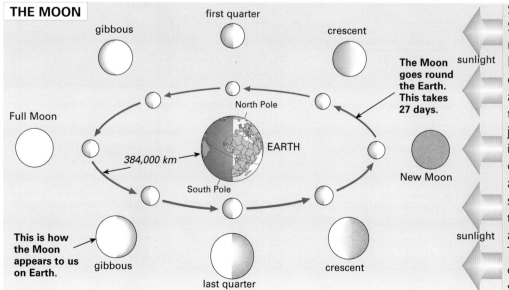

first quarter
gibbous
crescent
Full Moon
North Pole
EARTH
384,000 km
South Pole
New Moon
The Moon goes round the Earth. This takes 27 days.
sunlight
This is how the Moon appears to us on Earth.
gibbous
last quarter
crescent
sunlight

Some planets of the Solar System have satellites that revolve around them. The Earth has just one satellite, called the Moon. The Moon is about a quarter of the size of the Earth. It orbits the Earth in just over 27 days. The Moon is round but we on Earth see only the parts lit by the Sun and we never see 'the dark side'. This makes it look as if the Moon is a different shape at different times of the month. These are known as the phases of the Moon. Phases are shown in this diagram.

- Seasons happen because the Earth's axis is tilted at an angle of 23½°. The Earth revolves around the sun. This gives us the seasons of the year.
- In June, the northern hemisphere is tilted towards the Sun. As a result, it receives more hours of sunshine in a day and therefore has its warmest season, summer.
- Six months later, in December, the Earth has moved halfway round the Sun so that the southern hemisphere is tilted towards the Sun and it has its summer.

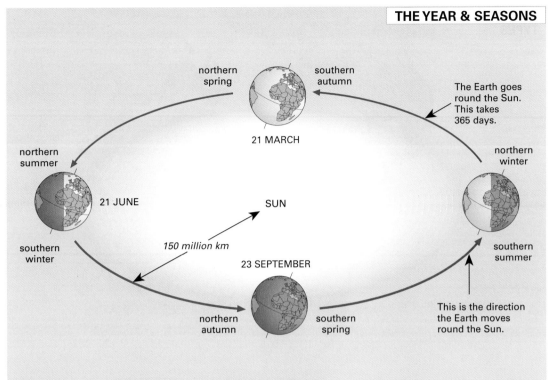

THE YEAR & SEASONS

northern spring

southern autumn

21 MARCH

The Earth goes round the Sun. This takes 365 days.

northern summer

21 JUNE

SUN

150 million km

southern winter

23 SEPTEMBER

northern autumn

southern spring

northern winter

southern summer

This is the direction the Earth moves round the Sun.

Spring

Summer

Winter

Autumn

HOW THE LANDSCAPE RESPONDS TO THE SEASONS

In northern latitudes, the seasons each last about three months. Spring arrives in March, Summer in June, Autumn in September, and Winter in December. In the USA, Autumn is known as Fall, as leaves fall from trees in this season.

In southern latitudes (for example, the southern zone of South America, or Australia), the reverse is the case. Spring arrives in September, Summer in December, Autumn in March, and Winter in June.

DAY & NIGHT

- From Earth, the Sun appears to rise in the east, reach its highest point at noon, and then set in the west. In reality, it is not the Sun that is moving but the Earth has rotated from west to east.
- Due to the tilting of the Earth, the length of day and night varies. In June, the area above the Arctic Circle has constant daylight, while Antarctica above the Antarctic Circle has constant darkness. The situations are reversed in December. In the Tropics, the length of day and night varies little.

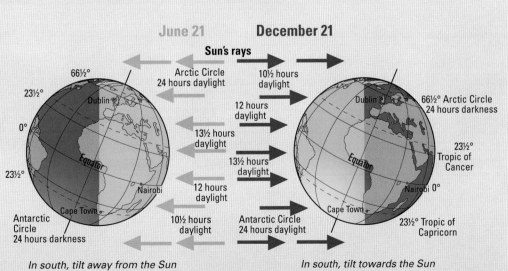

June 21   December 21

Sun's rays

66½°
Arctic Circle
24 hours daylight

23½°
Dublin

0°

Equator

23½°

Nairobi

Antarctic Circle
24 hours darkness

Cape Town

10½ hours daylight

12 hours daylight

13½ hours daylight

13½ hours daylight

12 hours daylight

10½ hours daylight

Dublin

66½° Arctic Circle
24 hours darkness

23½° Tropic of Cancer

Nairobi 0°

Equator

Cape Town

Antarctic Circle
24 hours daylight

23½° Tropic of Capricorn

In south, tilt away from the Sun

In south, tilt towards the Sun

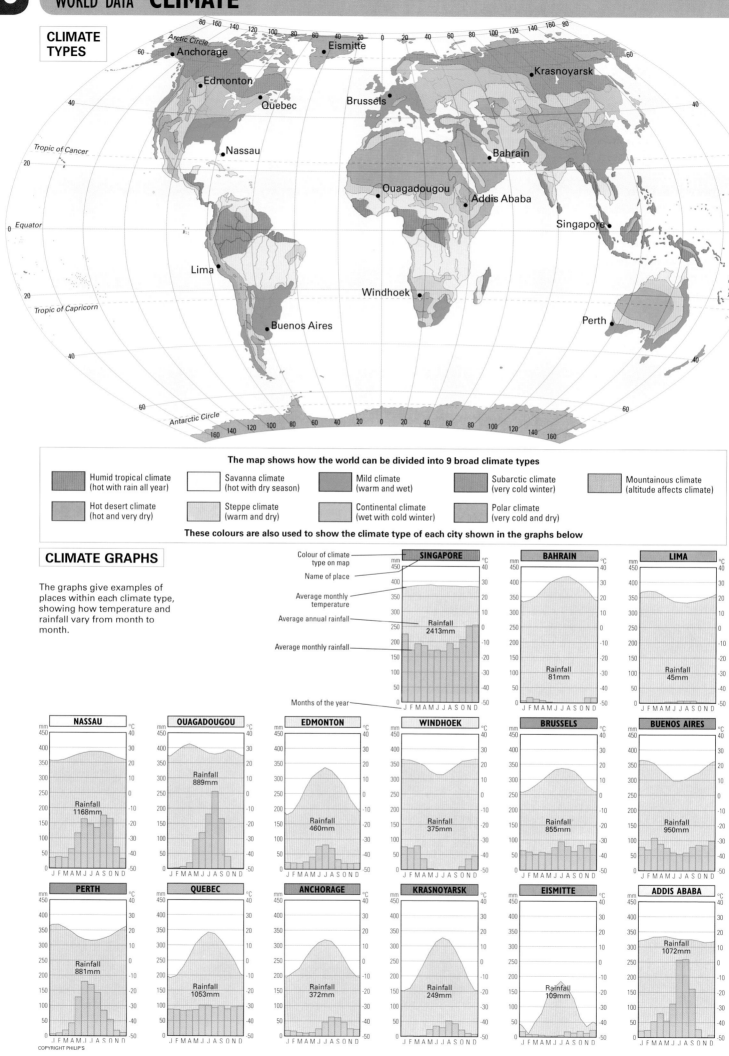

CLIMATE TYPES

The map shows how the world can be divided into 9 broad climate types

- Humid tropical climate (hot with rain all year)
- Hot desert climate (hot and very dry)
- Savanna climate (hot with dry season)
- Steppe climate (warm and dry)
- Mild climate (warm and wet)
- Continental climate (wet with cold winter)
- Subarctic climate (very cold winter)
- Polar climate (very cold and dry)
- Mountainous climate (altitude affects climate)

These colours are also used to show the climate type of each city shown in the graphs below

CLIMATE GRAPHS

The graphs give examples of places within each climate type, showing how temperature and rainfall vary from month to month.

Colour of climate type on map
Name of place
Average monthly temperature
Average annual rainfall
Average monthly rainfall
Months of the year

SINGAPORE — Rainfall 2413mm

BAHRAIN — Rainfall 81mm

LIMA — Rainfall 45mm

NASSAU — Rainfall 1168mm

OUAGADOUGOU — Rainfall 889mm

EDMONTON — Rainfall 460mm

WINDHOEK — Rainfall 375mm

BRUSSELS — Rainfall 855mm

BUENOS AIRES — Rainfall 950mm

PERTH — Rainfall 881mm

QUEBEC — Rainfall 1053mm

ANCHORAGE — Rainfall 372mm

KRASNOYARSK — Rainfall 249mm

EISMITTE — Rainfall 109mm

ADDIS ABABA — Rainfall 1072mm

COPYRIGHT PHILIP'S

Humid tropical climate

Hot desert climate

Savanna climate

Mild climate

Polar climate

Mountainous climate

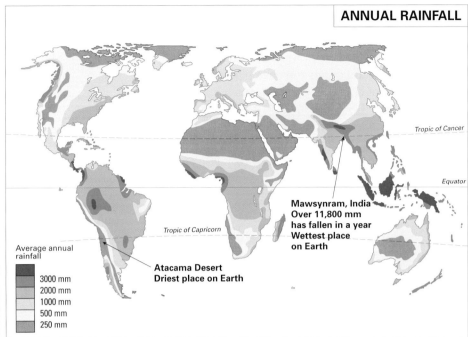

ANNUAL RAINFALL

Mawsynram, India
Over 11,800 mm
has fallen in a year
Wettest place
on Earth

Atacama Desert
Driest place on Earth

Average annual
rainfall

- 3000 mm
- 2000 mm
- 1000 mm
- 500 mm
- 250 mm

Tropic of Cancer

Equator

Tropic of Capricorn

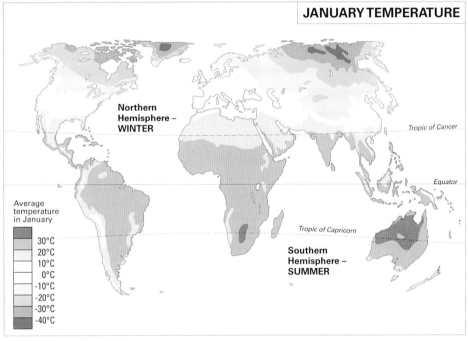

JANUARY TEMPERATURE

Northern
Hemisphere –
WINTER

Southern
Hemisphere –
SUMMER

Average
temperature
in January

- 30°C
- 20°C
- 10°C
- 0°C
- -10°C
- -20°C
- -30°C
- -40°C

Tropic of Cancer

Equator

Tropic of Capricorn

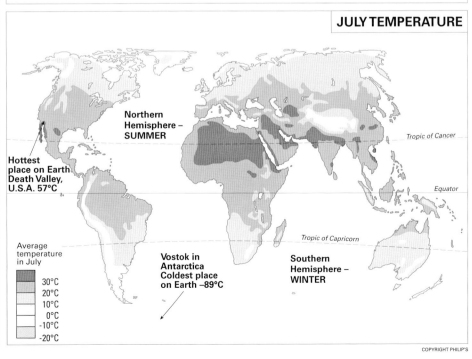

JULY TEMPERATURE

Northern
Hemisphere –
SUMMER

Hottest
place on Earth
Death Valley,
U.S.A. 57°C

Vostok in
Antarctica
Coldest place
on Earth –89°C

Southern
Hemisphere –
WINTER

Average
temperature
in July

- 30°C
- 20°C
- 10°C
- 0°C
- -10°C
- -20°C

Tropic of Cancer

Equator

Tropic of Capricorn

THE UNITED NATIONS

The United Nations organisation (UN) was established in 1945 to promote worldwide peace and co-operation. Its membership is 193 independent countries and its annual budget is around US$5 billion. Each member of the General Assembly has one vote. The Secretariat is the UN headquarters' administrative arm. The UN has 16 sectoral agencies, headquartered in the US, Canada, France, Switzerland, and other countries, which help members in sectors such as education (UNESCO), agriculture (FAO), health (WHO) and finance (IFC). The UN also has special offices such as those focussed on refugees (UNHCR) or on children (UNICEF). The Bahamas joined the UN on becoming an independent country in 1973.
• Can you find the purpose of all the agencies whose acronyms appear in this chart?

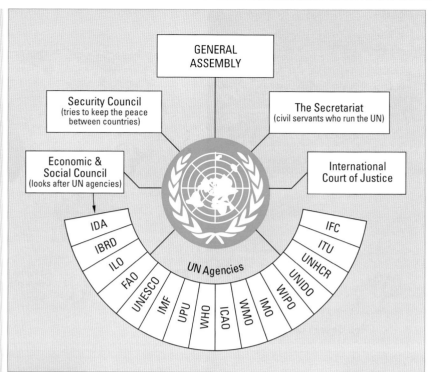

THE COMMONWEALTH

• The Commonwealth of Nations is an association of independent countries, most of which were once colonised by the British. Its objective is to strengthen democratic processes and to support development and economic growth.
• Several Commonwealth countries recognise the British monarch as their head of state. Most use English as one of their languages. There were 53 members in 2019, eleven of these in the Caribbean.
• UK Overseas Territories are also considered as part of the Commonwealth family. There are five such territories in the Caribbean, participating in Commonwealth activities.

COMMONWEALTH COUNTRIES: CARIBBEAN REGION

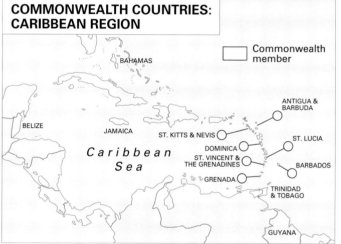

COMMONWEALTH COUNTRIES

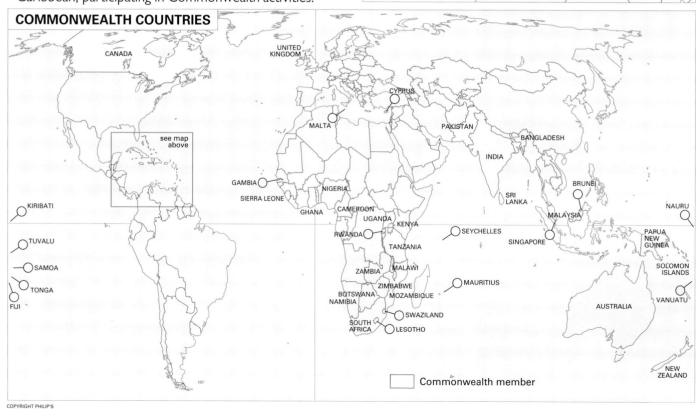

How the index is organised

All names in the index are listed in alphabetical order. Some names include a descriptor for the kind of feature it is (for example, the Gulf of Paria, to the west of Trinidad), the name is in alphabetical order followed by the description:

Paria, Gulf of

Sometimes, the same name occurs in more than one island or country. In these cases, the island or country names are added after each place name. For example:

Moss Town, *Cat I., Bahamas*
Moss Town, *Crooked I., Bahamas*

All river names are shown in blue (without the word river). For example, the Nile river in Egypt:

Nile

Every name in the index is followed by the page number of the map it appears on, a letter and then a number. For example:

Chaguanas **52** C2

The best map to find Chaguanas is on page 52. C2 is its grid reference. For a detailed explanation see page 6.

A lower case letter after the page number refers to the small map on that page.